TEXTS AND READINGS
IN MATHEMATICS **85**

Lie Groups and Lie Algebras

Lie Groups and Lie Algebras

M S Raghunathan

UM-DAE Centre for Excellence
in Basic Sciences
Mumbai

HINDUSTAN BOOK AGENCY

Published by

Hindustan Book Agency (India)
P 19 Green Park Extension
New Delhi 110 016
India

Email: info@hindbook.com
www.hindbook.com

ISBN 978-81-957829-5-6

For Ramaa

Contents

Preface

This is essentially a graduate text providing an introduction to Lie Groups and Lie Algebras. The book treats simultaneously real, complex and p-adic Lie Groups. Most available text-books on the subject deal with real Lie groups; one exception is Serre's 'Lie algebras and Lie Groups', which deals almost exclusively with p-adic Lie Groups. Also Lie algebras are treated much more extensively in this book than in most introductory text-books.

The first chapter outlines prerequisites, among them, some familiarity with p-adic fields and vector spaces over them (Chapter 1, X). It also fixes most of the notation in the book. The reader may skip this chapter and only refer to it when needed. The book assumes knowledge of undergraduate algebra and analysis; also the basic material on topology as is to be found in Munkres' book [7]. Some elementary notions and results about covering spaces are also outlined. A good reference for this material, as well as for results in Algebraic Topology, is Hatcher's book [3]. Some basic results about topological groups (as can be found e.g., Chapters 1–3 of Pontrjagin [9]) are stated without proof. The results not covered in most undergraduate courses are stated (and occasionally proved) in this first chapter. The results about local fields (stated without proof) can be found in Serre [12].

The second chapter is about analytic functions over locally compact fields. This chapter is elementary in character and simply carries over to a general, locally compact field with well-known definitions and results over real numbers. Many theorems are straightforward generalizations of statements in the real case; and the proofs of the results carry over to the non-archimedean case with minor modifications. However, the unfamiliar metric in the non-archimedean case could be a source of discomfiture for the student in translating the proofs in the real case to the case of non-archimedean fields. That is the reason for our dealing with this topic in some detail. It may be noted that the notion of C^k-functions does not work very well in the non-archimedean case.

Chapter 3 is on analytic manifolds. The material here again is a straightforward generalization of well-known elementary results on analytic manifolds over real numbers. However, in contrast to the real case, complex analysis or integration theory cannot be used to prove theorems about analytic functions or analytic manifolds over local fields. We do not touch upon the more subtle and deep aspects of the theory of real and complex analytic functions or real and complex analytic manifolds–they are not needed in our treatment of Lie groups.

Chapter 4 introduces Lie groups. The basic definitions such as those of the Lie algebra of a Lie group, the exponential map and Lie subgroups are made in this chapter, and some elementary results are proved. The methods are no different from those used to treat real Lie groups (in the existing text books).

The basic results viz., the theorem of E. Cartan that a closed subgroup of a Lie group is a Lie group and, the Fundamental Theorem of Lie Theory (due to Lie), which associates to each Lie subalgebra $\mathfrak{h}$ of the Lie algebra $\mathfrak{g}$ of a Lie group G, a Lie subgroup H, are proved in Chapter 5. Some proofs here are different from what one finds in standard books on Lie theory.

The sixth chapter embarks on the study of Lie algebras over an arbitrary field of characteristic zero. The basic theorem of Engels on Lie algebras of nilpotent matrices and the theorem of Lie on solvable Lie algebras are proved here. Cartan's criterion for solvability is also proved in this chapter: the proof is somewhat different from Cartan's proof which uses the theory of 'replicas'.

In the seventh chapter we prove the basic result about semi-simple Lie algebras, namely, that their finite dimensional representations are completely reducible. This holds the key to proving the structure theorem which says that a general Lie algebra $\mathfrak{g}$ is the semi-direct product of a semi-simple subalgebra $\mathfrak{s}$ and the solvable radical $\mathfrak{r}$ of $\mathfrak{g}$. We prove here also the Jacobson–Morozov theorem about embedding $\mathfrak{sl}(2, k)$ in a semi-simple Lie algebra. The book ends with a proof of Ado's theorem that any Lie algebra over a field k of characteristic 0 is isomorphic to a subalgebra of $M(n, k)$ for some n. The treatment follows essentially that of Bourbaki [2].

When k is a local field, theorems about a Lie algebra $\mathfrak{g}$ over k have implications for a Lie group G with $\mathfrak{g}$ as its Lie algebra, mostly through Lie's theorem. We do draw the attention of the reader to such implications in Chapters 6 and 7, but this is by no means exhaustive.

This book has taken several years for me to write, mainly because there were long intervals when I was not at the job. Also I kept revising parts already written repeatedly. Even now there are parts that can be written differently and better, but I do not want to further delay redeeming my promise to the publisher HBA, in particular, to Rajendra Bhatia, their editor for mathematics books. I am indeed thankful to them for putting up with this long delay.

We make free use of the Gothic script for notational purposes. For the convenience of the reader who is not familiar with the Gothic script we have at the end of the book given the English alphabet in both the Roman and Gothic scripts.

I am very grateful to T N Venkataramana, P Sankaran and A Khare who read through two earlier versions of the book diligently, pointed out numerous typographical errors, and made suggestions for improving the exposition, most of which, I have incorporated. The book has indeed become much more readable than the versions submitted to HBA earlier, thanks to their intervention.

Mumbai, July 1, 2023   M S Raghunathan

1. Notational Conventions and Other Preliminaries

We set down in this chapter some notational conventions (much of it standard) for the entire book. We also record here some facts and results from algebra and topology and about locally compact groups and fields without proofs which we will be using freely. The material described can be found in standard under-graduate/graduate texts.

I. Notation (General)

1.1. ϕ denotes the empty set.

For a set S, 1_S is the identity map of S on itself. For sets S, S', $f : S \to S'$ means a map f of S in S'.

For a set $\{S_i| \ i \in I\}$ of sets, $\coprod_{i \in I} S_i$ is the *disjoint* union of the S_i.

For a collection $\{S_i| \ i \in I\}$ of sets, $\prod_{i \in I} S_i$ is the product of the S_i, i.e., the set $\{f : I \to \coprod_{i \in I} S_i| \ f(i) \in S_i\}$.

For sets S, I, S^I (or $\prod_{i \in I} S$) denotes the set of all maps of I in S.

Note: S^ϕ is the singleton set consisting of the unique inclusion $\phi \subset S$.

For a finite set S, $|S|$ is the number of elements in S.

$\mathbb{Z}$ is the ring of integers and $\mathbb{Z}^+$ the subset of positive integers.

$\mathbb{N} \ (\supset \mathbb{Z}^+)$, the sub-semi-group of *non-negative* integers in $\mathbb{Z}$.

Note: This is not 'standard' notation: $\mathbb{N}$ is more often used to denote the set of *positive* ('natural') integers; we have included 0 in $\mathbb{N}$.

For integers $m, n, m \leq n$, $[m, n] = \{x \in \mathbb{Z}| \ m \leq x \leq n\}$.

For a set S and an integer $n \in \mathbb{N}$, we denote $S^{[1,n]}$, the n-fold product of S with itself; it is also denoted S^n.

The binary operation in a semi-group S is denoted '$\cdot$', except when it is commutative, in which case it is often denoted '$+$'.

'1' or 'e' denotes the identity in S if the binary operation is '$\cdot$'.

If the binary operation is denoted '$+$', the identity is denoted 0.

Let S be a set and $s, s' \in S$; then $\delta_{s,s'} = 0$ if $s \neq s'$ and $= 1$ if $s = s'$.

In a ring A, the addition is denoted '$+$' and the multiplication by '$\cdot$'; the additive identity is denoted 0 and the multiplicative identity 1.

1.2. $\mathbb{Q}$ will denote the field of rational numbers.

For $n \in \mathbb{N}$, $\mathbb{N}^n$ is a sub-semi-group of the abelian group $\mathbb{Z}^n$.

For $1 \leq i,\ j \leq n$, $\delta_{i,j} = 0$ if $i \neq j$ and $= 1$ if $i = j$.

For $i \in \mathbb{N}$, $<i>$ in $\mathbb{N}^n$ is the element $\{\delta_{i,j}\}_{1 \leq j \leq n}$.

For $n \in \mathbb{N}$, $n! = \prod_{i=1}^{n} i$ if $n \geq 1$ and $0! = 1$.

For $\alpha = \{\alpha_i\}_{1 \leq i \leq n} \in \mathbb{N}^n$, $|\alpha| = \sum_{1 \leq i \leq n} \alpha_i$, $\alpha! = \prod_{1 \leq i \leq n} \alpha_i!$.

For $\alpha, \beta \in \mathbb{N}^n$, $\alpha \leq \beta$ iff $\alpha_i \leq \beta_i$ for all i with $1 \leq i \leq n$.

For $a \in A$, A a ring containing $\mathbb{Q}$, and $r \in \mathbb{N}$, $\binom{a}{r} = (\prod_{i=0}^{(r-1)} (a - i))/r!$.

For a, r as above, we denote $\binom{a}{r}$ also by ${}^aC_r$.

If $a \in \mathbb{N}(\subset \mathbb{Q})$ and $r \in \mathbb{N}$ with $r \leq n$, evidently $\binom{a}{r} = a!/(r! \cdot (a-r)!)$.

For $\alpha, \beta \in \mathbb{N}^n$ with $\alpha \geq \beta$, $\binom{\alpha}{\beta} = \prod_{1 \leq i \leq n} \binom{\alpha_i}{\beta_i}$.

For $x = \{x_i\}_{1 \leq i \leq n} \in A^n$ (A a ring), and $\alpha \in \mathbb{N}^n$, $x^\alpha = \prod_{1 \leq i \leq n} x_i^{\alpha_i}$ where for $u \in A$, $u^0 = 1$.

1.3. $\mathbb{R}$ (resp. $\mathbb{C}$, resp. $\mathbb{H}$) denotes the field of real (resp. complex, resp. quaternionic) numbers.

Recall that the field $\mathbb{H}$ of quaternions is a 4-dimensional vector space over $\mathbb{R}$ with a basis $1, i, j, k$; and the multiplication in $\mathbb{H}$ is given by $1 \cdot x = x$ for all $x \in \mathbb{H}$, $i \cdot j = k = -j \cdot i$, $j \cdot k = i = -k \cdot j$ and $k \cdot i = j = -i \cdot k$.

For a quaternion $\alpha = a \cdot 1 + b \cdot i + c \cdot j + d \cdot k$, $\bar{\alpha} = a \cdot 1 - b \cdot i - c \cdot j - d \cdot k$ and the real part $\mathrm{Re}(\alpha)$ of α is a. One has $||\alpha|| \overset{\mathrm{def}}{=} \sqrt{\bar{\alpha} \cdot \alpha} = \sqrt{\alpha \cdot \bar{\alpha}} = \sqrt{a^2 + b^2 + c^2 + d^2}$; and $||\alpha|| > 0$ iff $\alpha \neq 0$.

The standard inner product on $\mathbb{R}^n$ (resp. Hermitian inner product on $\mathbb{C}^n$, resp. Quaternionic inner product on $\mathbb{H}^n$) is denoted $< \cdot, \cdot >$: for vectors $x = \{x_i\}_{1 \leq i \leq n}$ and $y = \{y_i\}_{1 \leq i \leq n}$, $<x, y> = \sum_{1 \leq i \leq n} \bar{x}_i \cdot y_i$. One has then $||x|| \overset{\mathrm{def}}{=} \sqrt{<x, x>} = \sqrt{\sum_{1 \leq i \leq n} ||x_i||^2}$.

For $x \in \mathbb{R}$, $[x]$ is the largest integer $\leq x$.

$\mathbb{R}^\times = \{x \in \mathbb{R} \,|\, x \neq 0\}$, $\mathbb{R}^{\geq 0} = \{x \in \mathbb{R} \,|\, x \geq 0\}$ and $\mathbb{R}^+ = \mathbb{R}^\times \cap \mathbb{R}^{\geq 0}$.

$\mathbb{C}^\times = \{x \in \mathbb{C} \,|\, x \neq 0\}$, $\mathbb{H}^\times = \{x \in \mathbb{H} \,|\, x \neq 0\}$.

$\mathbb{F}_p$ is the finite field $\mathbb{Z}/(p \cdot \mathbb{Z})$ ($p \in \mathbb{N}$ a prime) and for $q = p^n$, $\mathbb{F}_q$ is the finite field of q elements.

Unless otherwise specified, k will denote a field of characteristic 0.

II. Set Theory

1.4. Relations. A *relation* R on a set S is a subset $R \subset S \times S$.

1.5. A relation R on S is *reflexive*, if for every $x \in S$, $(x, x) \in R$.

R is *transitive*, if $(x, y) \in R$ and $(y, z) \in R$, implies that $(x, z) \in R$.

R is *symmetric*, if $(x, y) \in R$, implies that $(y, x) \in R$.

1.6. Order Relation. A relation R on S is an *order* relation, if it is reflexive and transitive. For an order relation R we write $x \leq_R y$ or $y \geq_R x$ iff $(x, y) \in R$. Evidently $x \leq_R x$ for all $x \in S$. We also write $x <_R y$ or $y >_R x$ for $(x, y) \in R$ if $x \neq y$. It follows that if $x \leq y$ and $y \leq x$ then $x = y$. That R is transitive means that for $x, y, z \in S$, $x \leq_R y$ and $y \leq_R z$ together imply that $x \leq_R z$. Two elements $x, y \in S$ are *comparable* for (the order relation) R, if either $x \leq_R y$ or $y \leq_R x$. When the order relation R used is clear from the context, we drop the suffix R and write simply $x \leq y$ for $x \leq_R y$.

1.7. An *ordered set* is a pair (S, R) where S is a set and R is an order relation on it.

1.8. Let (S, R) be an ordered set. Let T be a subset of S. Then $(T, R_T = R \cap (T \times T))$ is an ordered set and R_T is the *order on T induced by R*. An order R on a set S is a *total order* if any two elements in S are comparable. Let T be a subset of S. An upper bound for T in S is an element $u \in S$ such that $t \leq u$ for every $t \in T$. An element $m \in S$ is a *maximal* element for R if the following holds: if $m' \in S$ is such that $m' \geq m$, then $m' = m$. With these definitions we have the following basic result which is a very useful tool in many proofs.

1.9. Zorn's Lemma. Let (S, R) be an ordered set. Suppose that every subset T, such that R_T is a total order on T, admits an upper bound u_T in S. Then (S, R) admits a maximal element.

1.10. Equivalence Relation. A relation R on a set is an equivalence relation iff R is reflexive, symmetric and transitive. For an equivalence relation R on S, one writes $x \sim_R y$ to mean that $(x, y) \in R$ (and the suffix R is dropped if it is known from the context).

1.11. Let R be an equivalence relation on a set S. For $x \in S$, the R-*equivalence class of* $[x]$ is the set $\{y \in S | \ y \sim x\}$. From the definition of an equivalence relation, one sees easily that for $x, y \in S$, either $[x] = [y]$ or $[x] \cap [y] = \phi$. It follows that S is the disjoint union of equivalence classes. The set of equivalence classes is denoted $S/\sim$. One then has an obvious map $x \rightsquigarrow [x]$ of S in $S/\sim$.

1.12. $S/\sim_R$, the set of equivalence classes, is the *quotient of S by the equivalence relation R*.

III. Groups

1.13. Unless otherwise specified G will denote a group. For a group G and normal subgroups $G_1, G_2, [G_1, G_2]$ is the (normal) subgroup of G generated by $\{a \cdot b \cdot a^{-1} \cdot b^{-1} | \ a \in G_1, b \in G_2\}$ (note that $[G_1, G_2] = [G_2, G_1]$). $G^1 = [G, G]$ is the *commutator subgroup* of G.

For a group G, $A(G)$ denotes the automorphism group of G. Suppose that H is a group and $\Phi : H \to A(G)$ is a homomorphism. Then we can define a group structure on the set $H \times G$ as follows: $(h, x) \cdot (h', x') = (h \cdot h', \Phi(h'^{-1})(x) \cdot x')$. Then $H \times G$ equipped with this multiplication is the *semi-direct product of H and G (through Φ)* and is denoted $H \propto_\Phi G$. Note that $(H \simeq)H \times \{1\}$ and $(G \simeq)\{1\} \times G$ are subgroups of $H \propto_\Phi G$ and that $\{1\} \times G$ is a normal subgroup of $H \propto_\Phi G$. If Φ is the trivial homomorphism, $H \propto_\Phi G$ is the direct product of H and G.

1.14. Set $D^0(G) = G^0 = G$ and for integers $i > 0$, define $D^i(G)$ (also denoted G^i) inductively by setting $D^i(G) = G^i = [D^{(i-1)}(G), D^{(i-1)}(G)]$.
$\{D^i(G)| \ i \in \mathbb{N}\}$ is the *Derived Series* of G and $l_D(G) = \inf\{r \in \mathbb{N}| \ D^r(G) = D^{(r+1)}(G)\}$ is the *length of the derived series* of G. If $D^r(G) \neq D^{(r+1)}(G)$ for every $r \in \mathbb{N}$, $l_d(G) = \infty$. G is *solvable* iff $G^i = \{e\}$ for all large i and hence $l_D(G) < \infty$ and $G^{l_D(G)} = \{1\}$. The $D^i(G)$ are normal subgroups of G. The group G is *solvable* if $l_D(G) < \infty$ and $D^{l_D(G)}(G) = \{1\}$.

1.15. Define the groups $DC^i(G)$ inductively: $DC^0(G) = G$ and for $i > 0$, $DC^i(G) = [G, DC^{i-1}(G)]$. Then $\{DC^i(G)| \ i \in \mathbb{N}\}$ is the *descending central series* of G. The $DC^i(G)$ are normal subgroups of G. The *length $l_{DC}(G)$ of the descending central series* is defined as follows: if $DC^{(i+1)}(G) \neq DC^i(G)$ for all $i \in \mathbb{N}$, $l_{DC}(G) = \infty$; otherwise, $l_{DC}(G) = \inf\{i \in \mathbb{N}| \ DC^i(G) = DC^{(i+1)}(G)\}$. G is *nilpotent* if $DC^i(G) = \{1\}$ for all large i.

1.16. $Z(G)$ denotes the centre of G. The normal subgroups $AC_i(G)$ of G are defined inductively as follows: $AC_0(G) = \{e\}$; for $i \geq 0$, $AC_{(i+1)}(G) = p_i^{-1}(Z(G/AC_i(G)))$ where $p_i : G \to G/AC_i(G)$ is the natural map. $\{AC_i(G)| \ i \in N\}$ is the *ascending central series* of G. The length of the ascending central series of G (denoted $l_{AC}(G)$) is ∞ if $AC_i(G) \neq AC_{i+1}(G)$ for all i and equal to $\inf\{i \in \mathbb{N}| \ AC_i(G) = AC_{i+1}(G)\}$ otherwise (here, we have set $AC_{-1}(G) = \{e\}$). G is *nilpotent* if and only if $AC_i(G) = G$ for all large i.

1.17. We denote by $k[G]$ the group algebra of G over k. Recall that the group algebra is the vector space with G as a basis being the underlying abelian group, and the multiplication is defined as follows: for $x = \sum_{g \in G} a_g \cdot g$ and $y = \sum_{g \in G} b_g \cdot g$, $x \cdot y = \sum_{g \in G} c_g \cdot g$ where, for $g \in G$, $c_g = \sum_{h \in G} a_h \cdot b_{(h^{-1} \cdot g)}$. One defines the (injective) map $i_G : G \to k[G]$ of G in $k[G]$ by setting for $g \in G$ $i_G(g) = \sum_{h \in H} \delta_{g,h} \cdot h$; it is a group homomorphism of G into the group of units in $k[G]$. The proof of the following theorem is left as an exercise to the reader.

1.18. Theorem. *Given any homomorphism Φ of G into the multiplicative group of units in a k-algebra A, there is a unique algebra homomorphism $\tilde\Phi : k[G] \to A$ such that $\Phi = \tilde\Phi \circ i_G$. The diagonal inclusion $g \rightsquigarrow (g, g)$ of G in $G \times G$ induces an algebra homomorphism $\Delta : k[G] \to k[G \times G](\simeq k[G] \otimes_k k[G])$.*

IV. Rings and Modules; Linear Algebra

1.19. Rings and Modules. Let A be a ring with 1. Recall that a module M over A or an A-module is an abelian group M together with a map $\mu_M : A \times M \to M$ satisfying the following condtions:
(i) For $m \in M$, $\mu_M(1, m) = m$.
(ii) For $a, b \in A$ and $m \in M$, $\mu_M((a + b), m) = \mu_M(a, m) + \mu_M(b, m)$ and $\mu_M(a, (\mu_M(b, m))) = \mu_M(a \cdot b, m)$.
(iii) For $a \in A$ and $m, n \in M$, $\mu_M(a, (m + n)) = \mu_M(a, m) + \mu_M(a, n)$.
A with the multiplication $\mu : A \times A \to A$, is obviously an A-module.
For $a \in A$ and $m \in M$, we write $a \cdot m$ for $\mu(a, m)$. A module M over A is finitely generated if there is a finite set $S \subset M$ such that every $m \in M$ can be expressed as a linear combination $\sum_{m \in S} a_m \cdot m$ with $a_m \in A$. The set S is a *set of generators* of M.
An A-submodule N of an A-module M is a subgroup N of M such that $\mu(A \times N) \subset N$. Hence N is an A-module with $\mu_N = \mu_M|_{A \times N}$.
If M, N are two A-modules, a morphism $f : M \to N$ is a morphism of the underlying abelian groups which satisfies: $f(a \cdot m) = a \cdot f(m)$ for all $a \in A$ and $m \in M$.
Let M, N be A-modules. $\mathrm{Hom}_A(M, N)$, the set of A-module morphisms of M in N, is an abelian group under the addition defined as follows: for $f, f' \in \mathrm{Hom}_A(M, N)$ and $x \in M$, $(f + f')(x) = f(x) + f'(x)$.
$\mathrm{Hom}_A(M, M)$ is also denoted $\mathrm{End}_A(M)$; its elements are *endomorphisms* of M. $\mathrm{End}_A(M)$ is a ring with composition of endomorphisms as the multiplication. The set of invertible elements of $\mathrm{End}_A(M)$, denoted Aut_M or $GL_A(M)$, is evidently a group.
An element of $\mathrm{Aut}_A(M)$ is an *automorphism* of M.
The direct sum of a collection $\{M_i| \ 1 \leq i \leq r\}$ of A-modules is the abelian group $\prod_{i=1}^{r} M_i$ with the action of A given by: for $a \in A$ and $\{m_i\}_{1 \leq i \leq r} \in \prod_{i=1}^{r} M_i$, $a \cdot \{m_i\}_{1 \leq i \leq r} = \{a \cdot m_i\}_{1 \leq i \leq r}$.
A^r is a direct sum of r copies of A and $GL(r, A) = GL_A(A^r)$.

A basic result about finitely generated modules is the following:

1.20. Theorem. *Let $M \neq \{0\}$ be a finitely generated module over the unital ring A. Then M admits a maximal proper A-submodule $N \neq M$. In other words there is a submodule $N \neq M$ such that the following holds: if $N' \neq M$ is an A-submodule containing N, then $N' = N$.*

1.21. Proof. Let $\mathcal{M}$ be the collection of all A-submodules of M other than M. $\mathcal{M} \neq \phi$ since $\{0\} \in \mathcal{M}$. Set inclusion defines an order on $\mathcal{M}$. Let $\mathcal{T} \subset \mathcal{M}$ be a subset which is totally ordered under inclusion and let $T = \cup_{E \in \mathcal{T}} E$. Then it is easy to see that T is an A-submodule. We claim that $T \neq M$. To see this, let $\{m_i\}_{1 \leq i \leq r}$ be a set of generators for M. If $T = M$, $m_i \in T$ for every i and hence is in T_i for some $T_i \in \mathcal{M}$. As for any i, j with $1 \leq i, j \leq r$,

$T_i \subset T_j$ or $T_j \subset T_i$, we see that there is a $l \in [1, r]$ such that $T_i \subset T_l$ for $1 \leq i \leq r$. But this means that $T_l = M$, a contradiction to T_l being in $\mathcal{T}$. Thus T is an upper bound for $\mathcal{T}$. It follows from Zorn's Lemma (1.9) that $\mathcal{M}$ admits a maximal element – a maximal submodule $N \neq M$.

1.22. Vector Space. When the ring A is a field k, a module over A is a *vector space over k.*

1.23. Let V be a vector space over k. We denote by $\mathrm{End}_k(V)$ (or $\mathrm{End}(V)$ if k is known from the context) the k-algebra of all k linear maps of V into itself. It is also denoted $\mathfrak{gl}_k(V)$ (or $\mathfrak{gl}(V)$). $GL_k(V)$ (or $GL(V)$) denotes the group of invertible elements in $\mathrm{End}_k(V)$.

1.24. Let $B = \{e_i | \ 1 \leq i \leq r\}$ be an ordered basis of V over k. For $T \in \mathrm{End}_k(V)$, and $1 \leq i, j \leq r$, define $T_{i,j} \in k$ by setting

$$T(e_i) = \sum_{1 \leq j \leq r} T_{ji} \cdot e_j \tag{$*$}$$

Then $\{T_{i,j}\}_{1 \leq i,j \leq n}$ is the *matrix of T with respect to the ordered basis B.* This matrix is denoted T_B. The map $T \rightsquigarrow T_B$ sets up a k-algebra isomorphism of $\mathrm{End}_k(V)$ on the algebra of $M(r, k)$ of $r \times r$-matrices over k. Recall that the product $S \cdot T$ of matrices $S = \{S_{i,j}\}_{1 \leq i,j \leq r}$ and $T = \{T_{i,j}\}_{1 \leq i,j \leq r}$ is the matrix $\{(S \cdot T)_{i,j} = \sum_{1 \leq k \leq r} S_{i,k} \cdot T_{k,j}\}_{1 \leq i,j \leq r}$.

1.25. If B, B' are ordered bases of V, we have for $T \in \mathrm{End}_k(V)$,

$$T_{B'} = C_B^{-1} \cdot T_B \cdot C_B$$

where $C : V \rightarrow V$ is the endomorphism of V defined by $C(e_i') = e_i$ for $1 \leq i \leq n$. Inner conjugation by an invertible matrix $U \in M(r, k)$ is denoted $\mathrm{Int}(U)$: for $T \in M(r, k)$, $\mathrm{Int}(U)(T) = U \cdot T \cdot U^{-1}$.

1.26. Let K be a Galois extension of k. Let V be a vector space over k and $V_K = V \otimes_k K$. Then V gets identified with a subset of V_K through the injective map $V \ni v \rightsquigarrow v \otimes 1$. For a subspace W of V, the natural map $W_K = W \otimes_k K \rightarrow V \otimes_k K = V_K$ is an injection and enables us to identify W_K as a subspace of V_K. We denote by Γ the Galois group of K over k. We then have a natural action of Γ as k-linear automorphisms of V_K: for vectors of the form $v \otimes \alpha$, $v \in V$, $\alpha \in K$ and $\gamma \in \Gamma$, $\gamma(v \otimes \alpha) = v \otimes \gamma(\alpha)$. It is then clear that for any subspace W of V, the K-linear subspace W_K of V_K is Γ-stable. We state without proof the following result (which we will have occasion to use).

1.27. Proposition. *A K-linear subspace E of V_K is of the form W_K for some k-linear subspace W of V if and only if E is stable under the Galois group and $W = E \cap V$.*

1.28. For vector spaces M, N over k, $M \otimes_k N$ denotes the tensor product of M and N over k. We often write $M \otimes N$ for $M \otimes_k N$. $T^q M$ (or $\otimes^q M$) denotes the q-fold tensor product of M with itself (over k).

1.29. All rings (and k-algebras) considered will be assumed to be unital. A *k-algebra* is a ring A equipped with the structure of a vector space over k such that the map $x \rightsquigarrow x \cdot 1$ is an injective ring homomorphism into the centre of A; k is identified with a subring of A through this injection. A k-algebra is a vector space over k. The k-algebra A is *central* if k is the centre of A. Note that $\mathrm{End}_k(V)$ is a central algebra.

1.30. Let A, B be k-algebras. Then $A \otimes_k B$ has a natural k-algebra structure: for $a, a' \in A$ and $b, b' \in B$, $(a \otimes b) \cdot (a' \otimes b') = (a \cdot a') \otimes (b \cdot b')$. Evidently elements of A (identified with $A \otimes 1$) commute with elements of B (identified with $1 \otimes B$). It is an easy exercise to see that for a group G, $k[G \times G] \simeq k[G] \otimes_k k[G]$, and that $\Delta : k[G] \to k[G \times G]$ is an algebra homomorphism of $k[G]$ in $k[G] \otimes_k k[G]$ (see 1.17, 1.18).

1.31. A *graded algebra* over k is an algebra A over k together with a decomposition of $A = \coprod_{q \in \mathbb{N}} A^q$ as a direct sum of k-subspaces $\{A_q | \ q \in \mathbb{N}\}$ such that $k \subset A^0$ and $x \cdot y \in A_{(p+q)}$ if $x \in A^p$ and $y \in A^q$. A^q is the qth *graded component* of A. An element $\alpha \in A$ is *homogeneous of degree q* if it belongs to A_q. If we say that an element $x \in A$ has degree q, it means that x is homogeneous of degree q.

A graded algebra $A = \coprod_{q \in \mathbb{N}} A_q$ is *anti-commutative* iff $\alpha \cdot \beta = (-1)^{p \cdot q} \cdot \beta \cdot \alpha$ for all $\alpha \in A_p$ and $\beta \in A_q$. Note that this means $\alpha^2 = 0$ for all α of odd degree and that elements of even degree commute with all of A.

1.32. The Tensor Algebra. For a vector space M (over k). The q-fold tensor product of M (over k) with itself is denoted $T^q(M)$ or $\otimes^q M$. We also set $T^0(M) = k$. The direct sum $\coprod_{q \in \mathbb{N}} T^q(M)$ has a natural structure of an associative algebra over k with $T^q(M)$ as the qth graded component. The multiplication in $T(M)$ is defined by setting for $\{x_i | \ 1 \le i \le p\}$ and $\{y_j | \ 1 \le j \le q\}$,

$$(x_1 \otimes x_2 \cdots \otimes x_p) \cdot (y_1 \otimes y_2 \cdots \otimes y_q) = z_1 \otimes z_2 \cdots \otimes z_{p+q}$$

where $z_i = x_i$ for $1 \le i \le p$ and $z_{p+j} = y_j$ for $1 \le j \le q$.

1.33. The *exterior algebra* of M denoted $\bigwedge M$ is the quotient of $T(M)$ by the two-sided ideal generated by $\{x \otimes x | \ x \in M\}$. The image $\bigwedge^q M$ of $T^q M$ in $\bigwedge M$ under the natural map $T(M) \to \bigwedge M$ is the qth *exterior power* of M; one has $\bigwedge M = \coprod_{q \in \mathbb{N}} \bigwedge^q M$ and this direct sum decomposition makes

$\bigwedge M$ into a graded *anti-commutative* algebra over k. The natural map $M = T^1 M \to \bigwedge^1 M$ is an isomorphism. The multiplication in $\bigwedge M$ is denoted '$\wedge$'.

1.34. When M is of dimension r over k, and $\{e_i|\ 1 \le i \le r\}$ is an ordered basis of M over k, $\bigwedge^q M$ is a vector space of dimension $\binom{r}{q}$ having $\{e_{i_1} \wedge e_{i_2} \wedge \cdots \wedge e_{i_q}|\ 1 \le i_1 < i_2 < \cdots < i_q \le r\}$ as a basis. In particular $\bigwedge^r M \simeq k$, $\bigwedge^q M = 0$ if $q > r$ and $\bigwedge M$ is of dimension 2^r.

1.35. For a set S, $k[\{X_s|\ s \in S\}]$ is the polynomial algebra over k in variables $\{X_s|\ s \in S\}$. It can be made into a graded algebra by assigning a degree $d(s) \in \mathbb{N}$ to each $s \in S$ and defining the qth graded component of $k[\{X_s|\ s \in S\}]$ as the k-linear span of the *monomials* $\{\prod_{s \in S'} X_s^{r(s)}|\ S' \subset S\ finite,\ r : S' \to \mathbb{N},\ \sum_{s \in S'} d(s) \cdot r(s)) = q\}$. $k[\{X_s|s \in S\}]$ is also denoted $k[\{X_s\}_{s \in S}]$ in the sequel.

V. Lie Algebras

1.36. A *Lie algebra* over k is a vector space $\mathfrak{g}$ together with a k-bilinear map (called 'Bracket Operation') $[\cdot, \cdot] : \mathfrak{g} \times \mathfrak{g} \to \mathfrak{g}$ satisfying the following two conditions:
(i) for $X, Y \in \mathfrak{g}$, $[X, Y] = -[Y, X]$ and
(ii) for $X, Y, Z \in \mathfrak{g}$, $[[X, Y], Z] + [[Y, Z], X] + [[Z, X], Y] = 0$.
Equation (i) is equivalent to:
(i') $[X, X] = 0$ for every $X \in \mathfrak{g}$.
Equation (ii) is known as the Jacobi Identity.
Unless otherwise specified gothic lower case letters will denote Lie algebras. Also most Lie algebras considered will be finite dimensional.

1.37. Let V be a vector space. Set for $X, Y \in V$, $[X, Y] = 0$. Then evidently $(V, [\cdot, \cdot])$ is a Lie algebra. A Lie algebra $\mathfrak{g}$ is *abelian* if $[X, Y] = 0$ for every $X, Y \in \mathfrak{g}$.

1.38. An associative algebra A carries a natural structure of a Lie algebra: one defines for X, Y in A, $[X, Y] = X \cdot Y - Y \cdot X$. In particular $\text{End}_k(V)$ has a Lie algebra structure. $\text{End}_k(V)$ is also denoted $\mathfrak{gl}(V)$, especially when we treat it as a Lie algebra. As has been remarked, $\text{End}_k(k^n)$ has a natural identification with $M(n, k)$, the algebra of $(n \times n)$ matrices over k (through an ordered basis of V). We let $\mathfrak{gl}(n, k)$ also to stand for $M(n, k)$ especially, when we want to treat $M(n, k)$ as a Lie algebra.

1.39. A *Lie subalgebra* of a Lie algebra $\mathfrak{g}$ is a vector subspace $\mathfrak{h}$ of $\mathfrak{g}$ such that $[X, Y] \in \mathfrak{h}$ if X, Y are in $\mathfrak{h}$. We will often subalgebra mean Lie subalgebra. An *ideal* in a Lie algebra $\mathfrak{g}$ is a subalgebra $\mathfrak{h}$ such that $[X, Y] \in \mathfrak{h}$ if $X \in \mathfrak{g}$ and $Y \in \mathfrak{h}$. If $\mathfrak{h}, \mathfrak{h}'$ are ideals in $\mathfrak{g}$, so are $\mathfrak{h} \cap \mathfrak{h}'$ and $\mathfrak{h} + \mathfrak{h}' (= \{(X + X')|\ X \in \mathfrak{h}\ X' \in \mathfrak{h}'\})$; so is $[\mathfrak{h}, \mathfrak{h}'] (= [\mathfrak{h}', \mathfrak{h}])$, the k-linear span of $\{[X, Y]|\ X \in \mathfrak{h}, Y \in \mathfrak{h}'\}$.

1.40. Set $D^0(\mathfrak{g}) = \mathfrak{g}$; for integers $i > 0$, $D^i(\mathfrak{g})$ is defined inductively by $D^i(\mathfrak{g}) = [D^{(i-1)}\mathfrak{g}, D^{(i-1)}\mathfrak{g}]$. The $D^i(\mathfrak{g})$, $0 \le i < \infty$ are ideals in $\mathfrak{g}$. The *Derived Series* of the Lie algebra $\mathfrak{g}$ is the sequence $\{D^i(\mathfrak{g})| \ i \in \mathbb{N}\}$ of ideals. The length $l_D(\mathfrak{g})$ of the derived series of $\mathfrak{g}$ is defined as follows: if $D^i(\mathfrak{g}) \ne D^{(i+1)}(\mathfrak{g})$ for every $i \in \mathbb{N}$, $l_D(\mathfrak{g}) = \infty$; otherwise $l_D(\mathfrak{g}) = \inf\{i \in \mathbb{N}| \ D^i(\mathfrak{g}) = D^{(i+1)}(\mathfrak{g})\}$. A Lie algebra $\mathfrak{g}$ is *solvable* if $D^i(\mathfrak{g}) = \{0\}$ for $i >> 0$. Equivalently $l_D(\mathfrak{g}) < \infty$ and $D^{l_D(\mathfrak{g})}(\mathfrak{g}) = \{0\}$

1.41. The *Descending Central Series of* $\mathfrak{g}$ is the sequence of ideals $\{DC^i(\mathfrak{g})| \ i \in \mathbb{N}\}$ defined as follows: $DC^0(\mathfrak{g}) = D^0(\mathfrak{g}) = \mathfrak{g}$; for $i > 0$, $DC^i(\mathfrak{g}) = [\mathfrak{g}, DC^{i-1}(\mathfrak{g})]$. The *length $l_{DC}(\mathfrak{g})$ of the descending central series* of $\mathfrak{g}$ is ∞ if $DC_i(\mathfrak{g}) \ne DC_{i+1}(\mathfrak{g})$ for all $i \in \mathbb{N}$ and is $\inf\{i \in \mathbb{N}| \ DC^i(\mathfrak{g}) = DC^{i+1}(\mathfrak{g})\}$ otherwise. Note that $DC^i(\mathfrak{g}) \supset D^i(\mathfrak{g})$ for all i and $DC^i(\mathfrak{g}) = D^i(\mathfrak{g})$ for $i = 0, 1$. A Lie algebra $\mathfrak{g}$ is *nilpotent* if $DC^i(\mathfrak{g}) = \{0\}$ for all large i.

1.42. The *Ascending Central Series* of the Lie algebra $\mathfrak{g}$ is the sequence $\{AC_i(\mathfrak{g})| \ i \in \mathbb{N}\}$ of ideals defined as follows. $AC_0(\mathfrak{g}) = \mathfrak{c}(\mathfrak{g}) = \{X \in \mathfrak{g}| \ [X, Y] = 0, \ \forall Y \in \mathfrak{g}\}$, the *centre* of $\mathfrak{g}$. For $i \ge 1$, $AC_i(\mathfrak{g})$ is the inverse image of the centre of $\mathfrak{g}/AC_{(i-1)}(\mathfrak{g})$ in $\mathfrak{g}$ under the natural map $\mathfrak{g} \to \mathfrak{g}/AC_{(i-1)}(\mathfrak{g})$. The *ascending central series* of $\mathfrak{g}$ is the sequence $\{AC_i(\mathfrak{g})| \ i \in \mathbb{N}\}$ of ideals in $\mathfrak{g}$. A Lie algebra $\mathfrak{g}$ is nilpotent if and only if $AC_i(\mathfrak{g}) = \mathfrak{g}$ for all sufficiently large $i \in \mathbb{N}$.

1.43. Let $\mathfrak{g}$ be a Lie algebra over k and let $I_{\mathfrak{g}}$ be the two-sided ideal of the tensor algebra $T(\mathfrak{g})$ of $\mathfrak{g}$ generated by $\{X \otimes Y - Y \otimes X - [X, Y]| \ X, Y \in \mathfrak{g}\}$. The *(universal) enveloping algebra* $U(\mathfrak{g})$ of $\mathfrak{g}$ is the algebra $T(\mathfrak{g})/I_{\mathfrak{g}}$. This algebra together with the natural linear map $i = i_{\mathfrak{g}} : \mathfrak{g} \to U(\mathfrak{g})$ (obtained as the composite of the map $T(\mathfrak{g}) \to U(\mathfrak{g})$ with the inclusion of $\mathfrak{g}$ in $T(\mathfrak{g})$) has the following properties:
(i) $i([X, Y]) = i(X) \cdot i(Y) - i(Y) \cdot i(X)$.
(ii) If $u : \mathfrak{g} \to A$ is a k-linear map of $\mathfrak{g}$ into an associative k-algebra A such that $u([X, Y]) = u(X) \cdot u(Y) - u(Y) \cdot u(X)$, then there is a unique k-algebra homomorphism $\tilde{u} : U(\mathfrak{g}) \to A$ such that $u = \tilde{u} \circ i$.
(iii) If $f : \mathfrak{g} \to \mathfrak{h}$ is a Lie algebra homomorphism, $i \circ f$ induces an associative algebra homomorphism $U(f) : U(\mathfrak{g}) \to U(\mathfrak{h})$ extending $i \circ f$.
These assertions are easy exercises. In Chapter 7, we will show that i is injective. Consider now the associative algebra $U(\mathfrak{g}) \otimes_k U(\mathfrak{g})$: the multiplication is given by $(a \otimes b) \cdot (c \otimes d) = (a \cdot c) \otimes (b \cdot d)$ for $a, b, c, d \in U(\mathfrak{g})$. This algebra is naturally isomorphic to $U(\mathfrak{g} \times \mathfrak{g})$, the enveloping algebra of the direct product of $\mathfrak{g}$ with itself. The diagonal inclusion $\Delta : \mathfrak{g} \to \mathfrak{g} \times \mathfrak{g}$ being a Lie algebra morphism, extends to an associative algebra morphism (also denoted Δ): $\Delta : U(\mathfrak{g}) \to U(\mathfrak{g} \times \mathfrak{g}) \simeq U(\mathfrak{g}) \otimes_k U(\mathfrak{g})$. For $X \in \mathfrak{g}$, $\Delta(X) = X \otimes 1 + 1 \otimes X$. Δ is the *diagonal* homomorphism of $U(\mathfrak{g})$ in $U(\mathfrak{g}) \otimes_k U(\mathfrak{g})$ (see 1.17, 1.18).

1.44. Derivations. A *derivation* of a Lie algebra $\mathfrak{g}$ over k is a k-linear map $D : \mathfrak{g} \to \mathfrak{g}$ satisfying the following condition: $\forall\, X, Y \in \mathfrak{g}$,

$$D([X,Y]) = [D(X), Y] + [X, D(Y)].$$

A simple calculation shows that if D_1, D_2 are derivations of $\mathfrak{g}$ so is $[D_1, D_2](= D_1 \circ D_2 - D_2 \circ D_1)$. Thus the subset of $\mathrm{End}_k(\mathfrak{g})$ of all the derivations of $\mathfrak{g}$, is a Lie subalgebra (over k) of $\mathrm{End}_k(\mathfrak{g})$.

The Lie algebra of derivations of $\mathfrak{g}$ is denoted $\mathfrak{a}(\mathfrak{g})$. Let $\mathfrak{h}$ be a Lie algebra and $\rho : \mathfrak{h} \to \mathfrak{a}(\mathfrak{g})$ a Lie algebra morphism. Then one defines a Lie algebra structure on $\mathfrak{h} \oplus \mathfrak{g}$ as follows: for $X, Y \in \mathfrak{g}$ or $\mathfrak{h}$ the bracket operation is as in the respective Lie algebras. If $X \in \mathfrak{h}$ and $Y \in \mathfrak{g}$, $[X,Y] = \rho(X)(Y)$. That $\rho(X)$ is a derivation ensures that this bracket operation satisfies the Jacobi identity. The direct sum of $\mathfrak{h}$ and $\mathfrak{g}$ with this Lie algebra structure is the *semidirect product of* $\mathfrak{h}$ *and* $\mathfrak{g}$ and is denoted $\mathfrak{h} \propto_\rho \mathfrak{g}$. It is evident that $(\mathfrak{h} \simeq) \mathfrak{h} \times \{0\}$ and $(\mathfrak{g} \simeq) \{0\} \times \mathfrak{g}$ are subalgebras of $\mathfrak{h} \propto_\rho \mathfrak{g}$ and the latter is in fact an ideal in $\mathfrak{h} \propto_\rho \mathfrak{g}$. If $\rho = 0$, $\mathfrak{h} \propto_\rho \mathfrak{g}$ is the direct product of $\mathfrak{h}$ and $\mathfrak{g}$.

VI. Representations

1.45. For a vector space V over k, $\mathfrak{gl}(V)$, $\mathrm{End}_k(V)$ or simply $\mathrm{End}(V)$ (resp. $GL_k(V)$ or simply $GL(V)$) denotes the k-Lie algebra (resp. group) of all k-linear endomorphisms (resp. automorphisms) of V. $\mathrm{End}(V)$, as has been noted, has a natural structure of a Lie algebra given by: for $X, Y \in \mathrm{End}(V)$, $[X,Y] = X \cdot Y - Y \cdot X$. We denote $\mathrm{End}(V)$ by $\mathfrak{gl}_k(V)$ or $\mathfrak{gl}(V)$ especially when we want to treat it as a Lie algebra over k.

1.46. Let G (resp. $\mathfrak{g}$) be a group (resp. Lie algebra over k). Let K be a field extension of k. A *representation* ρ of G (resp. $\mathfrak{g}$) on a vector space V_ρ over K is a group morphism (resp. (k-linear) Lie algebra morphism) ρ of G (resp. $\mathfrak{g}$) into the group $GL_K(V_\rho)$ (resp. Lie algebra $\mathfrak{gl}_K(V_\rho)$). V_ρ is the *representation space* of ρ. We often let V_ρ stand for the representation ρ itself. Observe that a representation $\rho : G \to GL(V)$ (resp. $\mathfrak{g} \to \mathfrak{gl}(V)$) yields an associative k-algebra homomomorphism of $k[G]$ (resp. $U(\mathfrak{g})$) into $\mathrm{End}_K(V)$ extending ρ (see 1.18). Conversely such an associative algebra homomorphism composed with the natural map $G \to k[G]$ (resp. $\mathfrak{g} \to U(\mathfrak{g})$) gives a representation of G (resp. $\mathfrak{g}$) over K. Thus there is a bijective correspondence between representations of G (resp. $\mathfrak{g}$) over K and modules over the K-algebra $K[G] (\simeq k[G] \otimes_k K)$ (resp. $U(\mathfrak{g} \otimes_k K) (\simeq U(\mathfrak{g}) \otimes_k K)$). In the sequel we will use the term representations of G (resp. $\mathfrak{g}$) and $k[G]$-modules or G-modules (resp. $U(\mathfrak{g})$- modules or $\mathfrak{g}$-modules) interchangeably.

1.47. Examples. (1) The simplest example of a representation (over K) is the *trivial irreducible* representation of G (resp. $\mathfrak{g}$) (denoted $\underline{0}$). It is the representation on the 1-dimensional vector space K defined by setting for all $g \in G$ (resp. $X \in \mathfrak{g}$) $\underline{1}(g) = 1$ (resp. $\underline{0}(X) = 0$). A representation $\rho : \mathfrak{g} \to \mathfrak{gl}(V)$ is *trivial* iff $\rho(\mathfrak{g}) = \{0\}$.

(**2**) The identity map $1_{GL(V)}$ (resp. $1_{\mathfrak{gl}(V)}$) of $GL(V)$ (resp. $\mathfrak{gl}(V)$) into itself is evidently a representation (on V) of $GL(V)$ (resp. $\mathfrak{gl}(V)$) over k; it is called the *natural* representation of $GL(V)$ (resp. $\mathfrak{gl}(V)$) and is denoted ν. More generally if $G \subset GL(V)$ (resp. $\mathfrak{g} \subset \mathfrak{gl}(V)$) is a subgroup (resp. Lie subalgebra) ν restricted to G (resp. $\mathfrak{g}$) is the *natural* representation of G (resp. $\mathfrak{g}$) and is denoted ν_V.

1.48. Of particular interest is the *adjoint representation* $ad_{\mathfrak{g}} : \mathfrak{g} \to \mathfrak{gl}(\mathfrak{g})$ of a Lie algebra $\mathfrak{g}$. It is defined as follows: for $X \in \mathfrak{g}$, $ad_{\mathfrak{g}}(X)(Y) = [X, Y]$; the representation space $V_{ad_{\mathfrak{g}}}$ is evidently $\mathfrak{g}$. That $ad_{\mathfrak{g}}$ is a Lie algebra homomorphism is equivalent to the Jacobi identity. $ad_{\mathfrak{g}}$ is denoted ad when the Lie algebra $\mathfrak{g}$ is known in the context.

1.49. Let ρ, ρ' be representations of G (resp. $\mathfrak{g}$) over K. A *homomorphism* or *morphism* $f : \rho \to \rho'$ is a K-linear map $f : V_\rho \to V_{\rho'}$ such that $f(\rho(g)(v)) = \rho'(g)(f(v))$ for all $g \in G$ (resp. $f(\rho(X)(v)) = \rho'(X)(f(v))$ for all $X \in \mathfrak{g}$). In other words f is a morphism of G- (resp. $\mathfrak{g}$-) modules. The representations ρ and ρ' are *equivalent* if there is an isomorphism $f : V_\rho \to V_{\rho'}$ of G- (resp. $\mathfrak{g}$-) modules. A *sub-representation* of a representation ρ' of G (resp. $\mathfrak{g}$) is a pair $(\rho, f : \rho \to \rho')$ where ρ is a representation of G (resp. $\mathfrak{g}$) and f is an injective G- (resp. $\mathfrak{g}$-) module homomorphism. Evidently, $f(V_\rho)$ is a K-vector subspace of $V_{\rho'}$ stable under $\rho'(G)$ (resp. $\rho'(\mathfrak{g})$). Conversely a $\rho'(G)$-stable (resp. $\rho'(\mathfrak{g})$-stable) K-vector subspace W of $V_{\rho'}$ determines a sub-representation $(\rho, i : W \subset V_{\rho'})$ where $V_\rho = W$, i is the inclusion of W in $V_{\rho'}$ and ρ is defined by setting for $g \in G$ (resp. $X \in \mathfrak{g}$), $\rho(g) = \rho'(g)|_W$ (resp. $\rho(X) = \rho'(X)|_W$) (considered as a map into W). We often let W or ρ or V_ρ stand for $(\rho, i : V_\rho \subset V_{\rho'})$, if i is implicitly known in the context.

1.50. If ρ is a K-sub-representation of ρ' the *quotient* ρ'/ρ of ρ' by ρ is the K-representation defined as follows: $V_{\rho'/\rho} = V_{\rho'}/V_\rho$ and for $g \in G$ (resp. $X \in \mathfrak{g}$) and $v \in V_{\rho'}$, $(\rho'/\rho)(g)(v + V_\rho) = \rho'(g)(v) + V_\rho$ (resp. $(\rho'/\rho)(X)(v + V_\rho) = \rho'(X)(v) + V_\rho$). Observe that if $\Phi : \rho \to \rho'$ is a homomorphism, the kernel and image of Φ are sub-representations of ρ and ρ' respectively and image Φ is isomorphic to $\rho/kernel(\rho)$.

1.51. In the rest of this section on representations, we fix the field $K \supset k$ and all representations considered, unless otherwise specified, will be representations over K. If σ, τ are sub-representations of a representation ρ, identifying V_σ and V_τ as vector subspaces of V_ρ the vector subspace $(V_\sigma + V_\tau) \overset{\text{def}}{=} \{v + w|\ v \in V_\sigma,\ w \in V_\tau\}$ is evidently stable under $\rho(G)$ (resp. $\rho(\mathfrak{g})$) and is thus a sub-representation of ρ.

1.52. The direct sum $\coprod_{i \in I} \rho_i$ of a collection $\{\rho_i | i \in I\}$ of representations, is the representation defined as follows: $V_{\coprod_{i \in I} \rho_i} = \coprod_{i \in I} V_{\rho_i}$ and for $g \in G$ (resp. $X \in \mathfrak{g}$) and $v = \{v_i\}_{i \in I}$,

$$(\coprod_{i \in I} \rho_i)(g)(v) = \{\rho_i(g)(v_i)\}_{i \in I}$$
$$(\text{resp.}(\coprod_{i \in I} \rho_i)(X)(v) = \{\rho_i(X)(v_i)\}_{i \in I}).$$

1.53. A representation ρ of G (resp. $\mathfrak{g}$) is *irreducible* over K if V_ρ admits no proper vector subspace over K stable under $\rho(G)$ (resp. $\rho(\mathfrak{g})$). A representation ρ of G (resp. $\mathfrak{g}$) is *completely reducible* if it is equivalent to a direct sum of a collection of irreducible representations. The following result is known as Schur's Lemma:

Let ρ be an irreducible representation of a group G (resp. Lie algebra $\mathfrak{g}$) over an algebraically closed field K. Then if $T \in \text{End}_k(V_\rho)$ commutes with every element of $\rho(G)$ (resp. $\rho(\mathfrak{g})$), T is a scalar.

This is proved as follows. Let λ be an eigen-value of T. Then, since K is algebraically closed $W = \{v \in V_\rho | T(v) = \lambda \cdot v\}$ is a non-zero subspace of V_ρ stable under $\rho(G)$ (resp. $\rho(\mathfrak{g})$). Since ρ is irreducible, $W = V_\rho$ which implies that T is the scalar endomorphism λ.

1.54. Composition Series, Jordan-Hölder Series. Two representations ρ, σ are *equivalent* if there is an isomorphism of ρ on σ. In other words ρ and σ are isomorphic as $K[G]$- or $U(\mathfrak{g})$- modules. The equivalence class of a representation ρ is denoted $[\rho]$. Let ρ be a representation of a group G (resp. Lie algebra $\mathfrak{g}$). A *composition series* for ρ is a set $\{\rho_i | 0 \le i \le q\}$ of sub-representations of ρ such that
(i) $V_{\rho_0} = \{0\}$, $\rho_q = \rho$,
(ii) for $0 \le i \le (q-1)$, ρ_i is a sub-representation of $\rho_{(i+1)}$ and
(iii) $\tau_{(i+1)} = \rho_{(i+1)}/\rho_i$ is irreducible.
In general for a representation on an infinite dimensional vector space, a composition series may not exist. Note that for a finite dimensional representation, a composition series does exist but it need not be unique; however the number q as well as the collection $JH(\rho) = \{[\tau_i] | 1 \le i \le q\}$ of equivalence classes of irreducible representations is uniquely determined by the equivalence class of ρ. $JH(\rho)$ is the *Jordan-Hölder series* of ρ (or of the equivalence class $[\rho]$ of ρ).

1.55. Proposition. *The following conditions on a representation ρ of a group G or a Lie algebra $\mathfrak{g}$ are equivalent.*
(i) *ρ is a direct sum of irreducible sub-representations.*
(ii) *Let $\mathcal{I}(\rho) = \{\sigma | \sigma \subset \rho, \sigma \text{ irreducible}\}$; then $E = \sum_{\rho \in \mathcal{I}(\rho)} \rho$.*
(iii) *If $\sigma \subset \rho$, there is a $\sigma' \subset \rho$ such that $\rho \simeq \sigma \oplus \sigma'$.*

1.56. Proof. That (i)$\Rightarrow$(ii) is obvious.

Next we show that (ii)$\Rightarrow$(iii). Let σ be a sub-representation of ρ. If $V_\sigma = \{0\}$ or V_ρ, there is nothing to prove. Assume then that $V_\sigma \neq \{0\}$, V_ρ. Let $\mathcal{I}(\rho)$ be the set of all irreducible sub-representations of ρ and let $\mathcal{F} = \{S \subset \mathcal{I}(\rho)|\ W = \sum_{\tau \in S} V_\tau$ is a direct sum and $V_\sigma \cap W = \{0\}\}$. Then $\mathcal{F}$ is non-empty: since σ is a proper sub-representation and V_ρ is the sum of all its irreducible sub-representations, there is an irreducible representation τ such that V_τ is not a subspace of V_σ; it follows from the irreducibility of τ that $V_\sigma \cap V_\tau = \{0\}$ and hence $\{\tau\} \in \mathcal{F}$. Inclusion gives a partial order on $\mathcal{F}$. If $\{S_i|\ i \in I\}$ is a totally ordered subset of $\mathcal{F}$, $\tilde{S} = \cup_{i \in I} S_i$ is evidently an upper bound for the totally ordered subset; and as is easily seen, it is in $\mathcal{F}$. It follows by Zorn's lemma (cf. 1.9) that $\mathcal{F}$ has a maximal element T. Let $V' = \sum_{\tau \in T} V_\tau$ – it is necessarily a direct sum. We claim that $V_\sigma \oplus V' = V_\rho$. If not, there exists a $\lambda \in \mathcal{I}(\rho)$ such that $V_\lambda \cap (V' \oplus V_\sigma) = \{0\}$; but this means that the sum $V' + V_\lambda$ is direct and $(V' \oplus V_\lambda) \cap V_\sigma = \{0\}$ contradicting the maximality T. Hence (ii)$\Rightarrow$(iii).

Assume now that (iii) holds. We will first show that ρ admits a non-zero irreducible sub-representation. To see this let $v \in V_\rho$ be any non-zero vector. Let M be the G- (or $\mathfrak{g}$-) submodule of V_ρ generated by v. Then by Zorn's lemma (cf. 1.9) M admits a maximal submodule $M_o \neq M$; and the representation on M/M_o is irreducible. If $M_o = \{0\}$, M is irreducible. If not let E be a submodule of V_ρ such that $M_o \oplus E = V_\rho$. Then $M = M_o \oplus (M \cap E)$; and $M \cap E$ is a non-zero irreducible submodule of V_ρ. Thus the set $\mathcal{I}(\rho)$ of (non-zero) irreducible sub-representations of ρ is non-empty. Let $V' = \sum_{\tau \in \mathcal{I}(\rho)} V_\tau$. Then V' is a sub-representation of ρ. We need to show that $V' = V$. If not there is a (non-zero) sub-representation ρ' of ρ such that $V = V' \oplus V_{\rho'}$ and $V_{\rho'}$ admits an irreducible sub-representation τ; but then $\tau \in \mathcal{I}(\rho)$, a contradiction. Hence (iii)$\Rightarrow$(i), completing the proof of the theorem.

1.57. A representation ρ over k defines for any extension field K of k, a representation ρ_K over K, viz., the composition of the inclusion $GL_k(V_\rho) \subset GL_K(V_\rho \otimes_k K)$ (resp. $\mathfrak{gl}_k(V_\rho) \subset \mathfrak{gl}_K(V_\rho \otimes_k K)$) with ρ. In view of 1.27 we have the following.

1.58. Proposition. *Let $\bar{k}$ be an algebraic closure of k. Then a representation ρ (of G or $\mathfrak{g}$) over k is completely reducible if and only if $\rho_{\bar{k}}$ is completely reducible.*

1.59. Proof. We prove the proposition under the additional assumption that ρ is finite dimensional. Set $K = \bar{k}$. Let $\mathcal{G}$ be the Galois group of K over k. Suppose first that ρ is completely reducible. Since ρ is a direct sum of irreducible representations, we assume, as we may, that ρ is irreducible and that $V_\rho \neq \{0\}$. Let $\{0\} \neq E_K \subset V_{\rho_K}$ be any irreducible (over K) sub-representation of ρ_K. Then for any $\sigma \in \mathcal{G}$, $\sigma(E_K)$ is a K-vector subspace stable under the action of $\rho(G)$ (resp. $\rho(\mathfrak{g})$). It follows that $E'_K = \sum_{\sigma \in \mathcal{G}} \sigma(E_K)$

is a non-zero K-vector subspace of V_{ρ_K} stable under $\rho(G)$ (resp. $\rho(\mathfrak{g})$) as well as the Galois group Γ. It follows that there is a non-zero subspace E of V_ρ such that $E_K = E \otimes_k K$ (see 1.27). Evidently $E = E_K \cap V_\rho$ is non-zero and $\rho(G)$- (resp. $\rho(\mathfrak{g})$-) stable. Since ρ is irreducible $E = V_\rho$ and hence $V_K = E_K$ and since the $\sigma(E)$ are irreducible (over K) for every $\sigma \in \mathcal{G}$, V_{ρ_K} is completely reducible. Suppose now that ρ_K is completely reducible and $E \subset V_\rho$ is a sub-representation. Then $E_K (= E \otimes_k K)$ is a sub-representation of ρ_K. Since ρ_K is completely reducible, there is a K-linear map $p : V_{\rho_K} \to V_{\rho_K}$ such that
(i) $p(V_{\rho_k}) \subset E_K)$,
(ii) $p|_{E_K}$ is the identity 1_{E_K} of E_K and
(iii) $\rho_K(g) \circ p = p \circ \rho_K(g)$ resp. $\rho_K(X) \circ p = p \circ \rho_K(X)$ for all $g \in G$ (resp. $X \in \mathfrak{g}$). Now the orbit $\{\sigma^{-1} \circ p \circ \sigma | \sigma \in \mathcal{G}\}$ of p in $\mathrm{End}_K(V_{\rho_k})$ is of finite cardinality (= q say). Then $q^{-1}(\sum_{\sigma \in \mathcal{G}} \sigma^{-1} \circ p \circ \sigma)$ yields a projection of V_ρ on E compatible with the action of G (resp. $\mathfrak{g}$) on V_{ρ_K} and E_K. Thus ρ is completely reducible.

1.60. Let ρ, σ be representations over k. Then the representation $\mathrm{Hom}(\rho, \sigma)$ over K is defined as follows: the representation space is $\mathrm{Hom}_k(V_\rho, V_\sigma)$ and for $T \in \mathrm{Hom}_k(V_\rho, V_\sigma)$ and $g \in G$ (resp. $X \in \mathfrak{g}$), $\mathrm{Hom}(\rho, \sigma)(g)(T) = \sigma(g) \circ T \circ \rho(g)^{-1}$ (resp. $\mathrm{Hom}(\rho, \sigma)(g)(T) = \sigma(X) \circ T - T \circ \rho(X)$). When σ is the trivial (1-dimensional) representation $\mathrm{Hom}_k(\rho, \sigma)$ is the *dual* or *contragredient* representation of ρ. It is denoted ρ^*. If ρ is trivial $\mathrm{Hom}(\rho, \sigma)$ has a natural identification with σ.

1.61. Let ρ and σ be representations (over k). Then one defines the tensor product $\rho \otimes \sigma$ of ρ and σ (over k) by setting for $v \in V_\rho$, $w \in V_\sigma$ and $g \in G$ (resp. $X \in \mathfrak{g}$),

$$(\rho \otimes \sigma)(g)(v \otimes w) = \rho(g)(v) \otimes \sigma(g)(w) \text{ (resp.}$$
$$(\rho \otimes \sigma)(X)(v \otimes w) = \rho(X)(v) \otimes w + v \otimes \sigma(X)(w)).$$

An alternative description of the tensor product is as follows: V_ρ and V_σ may be regarded as modules over $K[G]$ (resp. $U(\mathfrak{g})$). It follows that $V_\rho \otimes_k V_\sigma$ has a natural structure of a module over $k[G] \otimes_k k[G]$ (resp. $U(\mathfrak{g}) \otimes_k U(\mathfrak{g})$) and hence composing with the diagonal map (1.18, 1.27) we obtain on $V_\rho \otimes_k V_\sigma$ a structure of a $k[G]$- (resp. $U(\mathfrak{g})$-) module; and this is the representation $\rho \otimes \sigma$. The natural map of $V_{\rho^*} \otimes_k V_\sigma$ on $\mathrm{Hom}_k(V_\rho, V_\sigma)$ gives an identification of $\rho^* \otimes \sigma$ with $\mathrm{Hom}(\rho, \sigma)$. The following fact is easy to prove.

1.62. Lemma. *For a representation ρ of G (resp. $\mathfrak{g}$), V_ρ^G (resp. $V_\rho^{\mathfrak{g}}$) is the space $\{v \in V_\rho | \rho(g)(v) = v, \forall g \in G\}$ (resp. $\{v \in V_\rho | \rho(X)(v) = 0, \forall X \in \mathfrak{g}\}$) of G- (resp. $\mathfrak{g}$-) invariants in V_ρ. With this notation, $\mathrm{Hom}_k(V_\rho, V_\sigma)^G$ (resp. $\mathrm{Hom}_k(V_\rho, V_\sigma)^{\mathfrak{g}}$) is precisely the space of all G- (resp. $\mathfrak{g}$-) module morphisms of V_ρ in V_σ.*

1.63. Invariant Bilinear Forms. A bilinear form B on V_ρ is an element of $V_{\rho^*} \otimes V_{\rho^*}$; and one checks that B is $\mathfrak{g}$ invariant if and only if the following holds: for $X \in \mathfrak{g}$ and $v, w \in V_\rho$,

$$B(\rho(X)(v), w) + B(w, \rho(X)(v)) = 0.$$

It is also easily checked that for any finite dimensional representation ρ of $\mathfrak{g}$, the symmetric bilinear form $(X, Y) \rightsquigarrow \mathrm{Trace}(\rho(X) \cdot \rho(Y))$ on $\mathfrak{g}$ denoted B_ρ is a $\mathfrak{g}$-invariant bilinear form on $\mathfrak{g}$ for the *adjoint representation*. Note that if $\{V_i \mid 1 \leq i \leq n\}$ are $\mathfrak{g}$-submodules with $V_1 = \{0\}$, $V_n = V_\rho$ and $V_i \subset V_{(i+1)}$, then $B_\rho = B_{(\coprod_{1 \leq i < n} \tau_i)}$ where τ_i is the representation $V_{(i+1)}/V_i$: in other words,
$\mathrm{trace}(\rho(X) \cdot \rho(Y)) = \sum_{1 \leq i \leq n} \mathrm{trace}(\tau_i(X) \cdot \tau_i(Y))$ for all $X, Y \in \mathfrak{g}$.

1.64. Tensor Powers of a Representation. For a representation ρ and integers $p, q \geq 0$ we denote by $\otimes^{p,q}\rho$ the representation on $(\otimes^p V_\rho) \otimes (\otimes^q V_{\rho^*})$, the space of p-times covariant and q-times contravariant tensors of V. Note that $\otimes^{0,0}$ is the trivial representation on k. Also $\otimes^{p,0}\rho$ is denoted $\otimes^p \rho$, so that $\otimes^{0,q}\rho = \otimes^q \rho^*$.

1.65. Symmetric and Exterior Powers. Recall that the permutation group S_n acts on the n-fold tensor product $\otimes^n V_\rho$ of V with itself by permuting the factors. For a vector space V over k denoted by $\Sigma_n V$ (resp. $A_n V$) the subspace $\{w \in \otimes^n V \mid \sigma(v) = v, \ \forall \ \sigma \in S_n\}$ (resp. $\{w \in \otimes^n V \mid \sigma(v) = \epsilon_\sigma \cdot v, \ \forall \ \sigma \in S_n\}$) of *symmetric* (resp. *anti-symmetric*) tensors in $\otimes^n V$: here ϵ_σ is the signature of σ. The quotient of $\otimes^n V$ by $A_n V$ (resp. $\Sigma_n V$) denoted $S^n V$ (resp. $\bigwedge^n V$) is the nth *symmetric* (resp. *exterior*) power of V. The natural map $\otimes^n V \to S^n V$ (resp. $\otimes^n V \to \bigwedge^n V$) restricted to $\Sigma_n V$ (resp. $A_n V$) is an isomorphism. The subspace of symmetric tensors $\Sigma_p V_\rho$ (resp. antisymmetric tensors $A_p V_\rho$) in $\otimes^p(V_\rho)$ is stable under $(\otimes^p \rho)(G)$ (resp. $(\otimes^p \rho)(\mathfrak{g})$) and is thus a sub-representation of $\otimes^p V_\rho$. The quotient of $\otimes^p V_\rho$ by $\Sigma(\otimes^p V_\rho)$ (resp. by $A(\otimes^p V_\rho)$) is the pth *exterior* power (resp. *symmetric* power) of ρ and is denoted $\bigwedge^p \rho$ (resp. $S^p \rho$). $\Sigma_p(V_\rho)$ (resp. $A_p(V_\rho)$) (maps isomorphically on $S^p \rho$ (resp. $\bigwedge^p \rho$).

VII. Topological Spaces

1.66. Unless otherwise specified, all topological spaces considered will be Hausdorff, locally compact, metrizable and countable at infinity. (A topological space X is *countable at infinity* if it is a countable union of compact subsets). This implies that the space is second countable and hence has a countable dense subset.

We state below some theorems that we will have occasion to use.

1.67. Theorem. *Let $(X, d : X \times X \to \mathbb{R}^{\geq 0})$ be a compact metric space and $\mathfrak{U} = \{U_i | \ i \in I\}$ an open covering of X. Then there is a constant $\delta = \delta(\mathfrak{U}) > 0$ such that for every $x \in X$, the open disc $D(x, \delta) = \{y \in X | \ d(x, y) < \delta\}$ of radius δ around x is contained in U_i for some $i \in I$.*

Such a δ is a *Lebesgue Number* of the covering $\mathfrak{U}$.

1.68. A collection $\mathfrak{U} = \{U_i | \ i \in I\}$ of subsets of a topological space X (we assume that $U_i \neq U_j$ for $i \neq j$) is *locally finite* if for every point $x \in X$ has a neighbourhood U_x such that the set $\{i \in I | \ U_x \cap U_i \neq \phi\}$ is finite. If $\mathfrak{U} = \{U_i | \ i \in I\}$ is locally finite, then for any compact subset K of X, the set $\{i \in I | \ K \cap U_i \neq \phi\}$ is finite. This condition is equivalent to the local finiteness of $\mathfrak{U}$ if X is locally compact. The collection $\mathfrak{U}$ is of order $\leq n$ (an integer ≥ 0) if the following holds: let $S \subset I$ be a set of cardinality q such that $\bigcap_{i \in S} U_i \neq \phi$; then $q \leq (n + 1)$. Evidently any open covering of order $n, n \in \mathbb{N}$ is locally finite.

1.69. Let $\mathfrak{U} = \{U_i | \ i \in I\}$ and $\mathfrak{V} = \{V_j | \ j \in J\}$ be coverings of X. Then $\mathfrak{V}$ is a *refinement* of $\mathfrak{U}$ if there is a *refinement map* $r : J \to I$ such that $V_j \subset U_{r(j)}$ for every $j \in J$. A topological space X is *paracompact* if every open covering of X admits an open locally finite refinement.

It is known that a metric space is paracompact.

1.70. Theorem. *A locally compact space is paracompact if and only if it is a disjoint union of open and closed subsets each of which is countable at infinity.*

Yet another fact that we will have occasion to use is as follows.

1.71. Theorem. *Let $\{U_i | \ i \in I\}$ be a locally finite open covering of a paracompact space. Then there is an open covering $\{V_i | 1 \in I\}$ such that the closure $\bar{V}_i$ of V_i is contained in U_i for every $i \in I$.*

The covering $\{V_i | \ i \in I\}$ as in the theorem is a *shrinking* of the covering $\{U_i | \ i \in I\}$.

1.72. A subset A of a topological space X is *nowhere dense* if its interior is empty; equivalently if its complement is dense in X. With this definition we have the following.

1.73. Baire Category Theorem. *Let X be a complete metric space or a locally compact space. Let $\{U_n | \ n \in \mathbb{N}\}$ be a sequence of open dense subsets of X. Then $\bigcap_{n \in N} U_n$ is dense in X. Equivalently a countable union of nowhere dense closed subsets of X is nowhere dense.*

VIII. Covering Spaces and the Fundamental Group

1.74. Let X be a topological space and x_o a point on it. A *loop at x_o in X* is a continuous map $\alpha : [0,1] \to X$ with $\alpha(0) = \alpha(1) = x_0$. Let $L(X, x_o)$ be the set of all loops at x_o. Two loops α, β are *homotopic* (written $\alpha \sim \beta$) if there is a continuous map $H : I \times I \to X$ such that $H(I \times \{0\}) = H(I \times \{1\}) = x_o$ and $H(0,t) = \alpha(t)$ and $H(1,t) = \beta(t)$ for all $t \in [0,1]$. One has a composition $\alpha \cdot \beta$ of two loops α and β defined as follows: by $(\alpha \cdot \beta)(t) = \alpha(t)$ for $0 \le t \le 1/2$ and $\beta(2 \cdot t - 1)$ for $1/2 \le t \le 1$. Then $\sim$ is an equivalence relation on $L(X, x_o)$ compatible with the composition on it defined above; hence it defines a composition law on the set $\pi_1(X, x_o)$ of equivalence classes in $L(X, x_o)$. Under this composition law $\pi_1(X, x_o)$ is a group, the *fundamental group* of X at x_o. When X is arc-wise connected, its isomorphism class is independent of the point $x_o \in X$. A topological space X is *simply connected* if it is arc-wise connected and $\pi_1(X, x_o)$ is trivial.

1.75. A continuous map $f : X \to Y$ of topological spaces is a *local homeomorphism* if for every $x \in X$ there is an open set $U \ni x$ of such that $f|_U$ is a homeomorphism of U on to an open set $V = f(U)$ of Y. f is a *covering projection* if $f(X) = Y$ and for every point $y \in Y$ there is an open neighbourhood $U \ni y$ such that $f^{-1}(U)$ is a disjoint union $\coprod_{i \in I} U_i$ of open sets in X such that $f|_{U_i} : U_i \to U$ is a homeomorphism. A covering projection is necessarily a local homeomorphism.

1.76. Let Y be a connected locally contractible topological space, y_o a point on it and $\Gamma = \pi_1(Y, y_o)$ its fundamental group at y_o. Then *there is a connected, simply connected topological space $\tilde{Y}$, a point $\tilde{y}_o$ on it and a covering projection $u : \tilde{Y} \to Y$ with the following properties:*
(i) $u(\tilde{y}_o) = y_o$ and
(ii) *If $f : Y' \to Y$ is a covering projection and $y'_o \in Y'$ is a point such that $f(y'_o) = y_o$, there is a unique covering map $\tilde{f} : \tilde{Y} \to Y'$ such that $\tilde{f}(\tilde{y}_o) = y'_o$ and $u = f \circ \tilde{f}$.*
Evidently such a triple $(\tilde{Y}, \tilde{y}_o, u : \tilde{Y} \to Y)$ is unique up to an 'isomorphism' in an obvious sense and is called a *universal covering* of Y. A covering $\pi : X \to Y$ is a universal covering if and only if X is simply connected.

1.77. One has a more general property of the universal covering than what is stated above. *Let $q : Z' \to Z$ be a covering projection $z'_o \in Z'$ with Z and Z' connected and locally arc-wise connected. Let $f : Y \to Z$ a continuous map such that $f(y_o) = z_o = q(z'_o)$. Then there is a unique continuous map $\tilde{f} : \tilde{Y} \to Z'$ with $\tilde{f}(\tilde{y}_o) = z'_o$ and $q \circ \tilde{f} = f \circ u$.*

If $(\tilde{X}, \tilde{x}_o, u_X : \tilde{X} \to X)$ and $(\tilde{Y}, \tilde{y}_o, u_Y : \tilde{Y} \to Y)$ are universal coverings of (X, x_o) and (Y, y_o) respectively then $(\tilde{X} \times \tilde{Y}, (\tilde{x}_o, \tilde{y}_o), u_X \times u_Y : \tilde{X} \times \tilde{Y} \to X \times Y)$ is a universal covering of $X \times Y$.

We will be using the following result (see Hatcher [3] or Munkres [7]).

1.78. Theorem. *Let X be a connected paracompact locally contractible simply connected topological space. Let $\mathcal{U} = \{U_i| \ i \in I\}$ be an open covering of X. Suppose that we are given a collection $\{c_{i,j} \in \mathbb{R}| \ (i,j) \in I \times I, \ U_i \cap U_j \neq \phi\}$ of real numbers such that $c_{i,j} = -c_{j,i}$ and $c_{i,j} + c_{j,k} + c_{k,i} = 0$ for $i, j, k \in I$ with $U_i \cap U_j \cap U_k \neq \phi$, then there is an open refinement $(\mathcal{V} = \{V_j| \ j \in J\}, r : J \to I)$ of $\mathfrak{U}$ and elements $\{b_j| \ j \in J\}$ in $\mathbb{R}$ such that $b_j - b_i = c_{r(j),r(i)}$ for all $i, j \in J$.*

IX. Topological Groups

1.79. Recall that a *topological group* is a group G together with a Hausdorff topology on the underlying set of G such that the multiplication map $m : G \times G \to G$ and the map $\mathcal{I}v : G \to G$ given by $\mathcal{I}v(x) = x^{-1}$ are continuous. We will assume that the underlying topological space is locally compact, metrizable and second countable.

1.80. If G is a topological group and H a closed subgroup, the set G/H of right cosets $\{g \cdot H| \gamma \in G\}$ equipped with the quotient topology is a Hausdorff topological space such that the natural map $m : G \times G/H \to G/H$ defined by setting for $g \in G$ and $x \cdot H \in G/H$, $m(g, x \cdot H) = (g \cdot x) \cdot H$, is continuous. If H is a normal in G, m descends to the group multiplication in G/H making it into a topological group.

1.81. Let G be a connected locally contractible topological group G and $(\tilde{G}, u : \tilde{G} \to G)$ be a universal covering of G. Then the group multiplication $m : G \times G \to G$, lifts to a map $\tilde{m} : \tilde{G} \times \tilde{G} \to \tilde{G}$ with $\tilde{m}(\tilde{e}, \tilde{e}) = \tilde{e}$ where $\tilde{e}$ is a point in $\tilde{G}$ with $u(\tilde{e}) = e$; a $\tilde{m}$ makes $\tilde{G}$ into a topological group with u as a (continuous) homomorphism. One has moreover a natural identification of kernel(u_G) with $\pi_1(G, e)$. Kernel(u_G) is a discrete normal subgroup of $\tilde{G}$. Since $\tilde{G}$ is connected, so is the inner conjugation orbit of any element of kernel(u_G) under G. As such an orbit lies in a discrete group, it is reduced to a point. It follows that $\pi_1(G, e) = $ Kernel(u_G) is central in $\tilde{G}$ and is hence abelian.

1.82. Let G be a locally compact topological group. For $g \in G$, let L_g (resp. R_g) denote the left (resp. right) translation by g: $L_g : G \to G$ (resp. $R_g : G \to G$) is the continuous map $G \ni x \rightsquigarrow g \cdot x$ (resp. $G \ni x \rightsquigarrow x \cdot g^{-1}$). Then we have the following.

1.83. Theorem. *Let G be a locally compact group. Then there is a Borel measure μ (resp. μ') on G with the following properties*:
(i) *For every Borel subset E of G, $\mu(L_g(E)) = \mu(E)$ (resp. $\mu'(R_g(E)) = \mu'(E)$) for every $g \in G$.*
(ii) *For every compact subset $K \subset G$, $\mu(K) < \infty$ (resp. $\mu'(K) < \infty$).*
(iii) *For a non-empty open set $U \subset G$, $\mu(U) > 0$ (resp. $\mu'(U) > 0$).*

The measure μ (resp. μ') is unique up to scaling.

A measure μ (resp. μ') as in the theorem is a left-(resp. right-) translation invariant *Haar measure* on G.

1.84. Corollary. *Let μ, μ' be as in the theorem above. Then for every g in G and a Borel set $E \subset G$, $\mu(R_g(E)) = \Delta(g) \cdot \mu(E)$, where $\Delta : G \to \mathbb{R}^+$ is a continuous homomorphism of G into the multiplicative group $\mathbb{R}^+$.*

Δ is called the *modular character* on G and G is said to be *unimodular* if Δ is trivial. Examples of unimodular groups are compact groups and groups G for which $[G, G]$ is dense in G.

X. Locally Compact Fields

A locally compact field is also referred to as a *local field*.

1.85. Absolute Value. Let $p \in \mathbb{N}$ be a prime. The *p-adic absolute value* on $\mathbb{Q}$ is the function $| \cdot |_p : \mathbb{Q} \to \mathbb{R}^{\geq 0}$ defined as follows: $|0|_p = 0$ and for $0 \neq x = p^r \cdot a/b$ where $a, b, r \in \mathbb{Z}$, a and b are co-prime to p and to each other; then $|x|_p = p^{-r}$. Evidently, one has $|x \cdot y|_p = |x|_p \cdot |y|_p$ for $x, y \in \mathbb{Q}$. This valuation is *non-archimedean* which means that it satisfies the following condition: for $x, y \in \mathbb{Q}$,

$$|x + y|_p \leq \max(|x|_p, |y|_p) \tag{$*$}$$

with equality holding if $|x|_p \neq |y|_p$.

1.86. Let $d_p : \mathbb{Q} \times \mathbb{Q} \to \mathbb{R}^{\geq 0}$ be the map defined by setting for $(x, y) \in \mathbb{Q} \times \mathbb{Q}$, $d_p(x, y) = |x - y|_p$; then d_p is a metric on $\mathbb{Q}$ making it into a topological field. The completion of $\mathbb{Q}_p$ of $\mathbb{Q}$ with respect to d_p is a locally compact field in which the closure $\mathbb{Z}_p$ of $\mathbb{Z}$ in is a maximal compact subring. The metric on $\mathbb{Q}$ extends uniquely to a metric on the completion $\mathbb{Q}_p$. It follows that the function $| \cdot |_p : \mathbb{Q} \to \mathbb{R}^{\geq 0}$ extends to a continuous function $| \cdot |_p : \mathbb{Q}_p \to \mathbb{R}^{\geq 0}$ satisfying $(*)$ for all $x, y \in \mathbb{Q}_p$.

1.87. Let k be a finite extension of $\mathbb{Q}_p$. Then k is a finite dimensional vector space (of dimension n, say) and hence can be identified (via an ordered basis of k over $\mathbb{Q}_P$) with $\mathbb{Q}_p^n$, the n-fold product of $\mathbb{Q}_p$ with itself. k equipped with the product topology on $\mathbb{Q}_p^n$, becomes a locally compact field. Recall that the norm $N_{k/\mathbb{Q}_p} : k \to \mathbb{Q}_p$ is the map $k \ni x \rightsquigarrow \det(m_x)$ where $m_x : k = \mathbb{Q}_p^n \to k = \mathbb{Q}_p^n$ is the multiplicaion by x. One defines an *absolute value* $|\cdot|_k$ on k as follows. For $x \in k$,

$$|x|_k = |N_{k/\mathbb{Q}_p}(x)|_p^{1/n}.$$

Note that n is the 'degree' of k over $\mathbb{Q}_p$. It is then clear that the restriction of $|\cdot|_k$ to $\mathbb{Q}_p$ equals $|\cdot|_p$. We also have for $x, y \in k$, $|(x+y)|_k \leq \max\{|x|_k, |y|_k\}$ with equality holding if $|x|_k \neq |y|_k$ so that $(x, y) \rightsquigarrow d_k(x, y) = |x - y|_k$ is a metric on k; the metric topology on k is the same as the topology got from identification with $\mathbb{Q}_p^n$.

1.88. Classification. The following theorem is known:
Any locally compact field k of characteristic 0 is isomorphic to either $\mathbb{R}$ or $\mathbb{C}$ or a finite extension k of $\mathbb{Q}_p$.
When $k \simeq \mathbb{R}$ or $\mathbb{C}$, $|x|_k$ for $x \in k$ will denote the usual absolute value (on $\mathbb{R}$ or $\mathbb{C}$ as the case may be). Fields isomorphic to $\mathbb{R}$ or $\mathbb{C}$ are *archimedean* while those that are not, are *non-archimedean*.

1.89. We denote by μ_k, the Haar measure on the additive grouip k normalized as follows: if k is non-archimedean, $\mu_k(\mathfrak{o}_k) = 1$ ($\mathfrak{o}_k$ is defined in 1.90 below). If $k = \mathbb{R}$, $\mu_k([0, 1]) = 1$. If $k = \mathbb{C}$, we take $\mu_{\mathbb{C}}$ to be $\mu_{\mathbb{R}} \times \mu_{\mathbb{R}}$ where we have identified $\mathbb{C}$ with $\mathbb{R}^2$ through the $\mathbb{R}$-basis $\{1, i\}$. On k^n we have the product measure μ_k^n which we denote μ_{k^n}.

1.90. Assume that k is a non-archimedean local field. The set $\mathfrak{o}_k = \{x \in k| \ |x|_k \leq 1\}$ is the *unique maximal compact subring* of k – it is the *ring of integers* in k. If p is the residue field characteristic, the closure of $\mathbb{Q}$ in k is (isomorphic) to $\mathbb{Q}_p$. Moreover $\mathfrak{o}_k$ contains the closure $\mathbb{Z}_p$ of $\mathbb{Z}$ in k and is a free module over $\mathbb{Z}_p$ of rank equal to $\dim_{\mathbb{Q}_p} k$. $\mathfrak{O}_k$ has a unique prime (hence also proper maximal) ideal which is denoted $\mathfrak{p}_k$. $\mathfrak{p}_k = \{x \in k| \ |x|_k < 1\}$. The quotient $F_k = \mathfrak{o}_k/\mathfrak{p}_k$ (called the *residue field* of k) is a finite field of p^{f_k} elements where p is the characteristic of F_k and f_k is the degree of F_k over the prime field $\mathbb{Z}/(p)$. The prime ideal $\mathfrak{p}_k$ is generated by an element π_k (which is unique up to multiplication by a unit in $\mathfrak{o}_k$). It is called a *uniformizing parameter*. Further every ideal in $\mathfrak{o}_k$ is a principal ideal generated by π_k^r for some integer $r > 0$. Hence $\mathfrak{o}_k$ is a principal ideal domain.

There is an integer $e_k \geq 1$ (called the *ramification index* of k over $\mathbb{Q}_p$) such that $e_k \cdot f_k = d_k$, the degree of the extension k over $\mathbb{Q}_p$. One has moreover $|\pi_k|_k = p^{-1/e_k}$. Also $p \cdot \mathfrak{o}_k = \mathfrak{p}_k^{e_k}$. Note that $|k^\times|_k$ is the *discrete* subgroup $\{p^{n/e_k}| \ n \in \mathbb{Z}\}$ of the multiplicative group $\mathbb{R}^\times$.

We set up some further notation that will be used freely in the book.

1.91. Notation. The closure of $\mathbb{Q}$ in k is denoted $\hat{\mathbb{Q}}$. It $\hat{\mathbb{Q}} \simeq \mathbb{R}$ if k is archimedean and to $\simeq \mathbb{Q}_p$, if k is non-archimedean with residue field characteristic p.

When k is non-archimedean, $\mathfrak{o}_k$ is the subring of *integers* in k: $\mathfrak{o}_k = \{x \in k \mid x \text{ is a root of}$

a monic polynomial in $\mathbb{Z}_p[X]$.

ϖ_k denotes the natural map of $\mathfrak{o}_k$ on F_k.

The *standard basis* of k^n is the basis $\{e_i = \{\delta_{i,j}\}_{1 \leq j \leq n} \mid 1 \leq i \leq n\}$.

For $x = \{x_i\}_{1 \leq i \leq n} \in k^n$, $||x||_k = \max\{|x_i| \mid 1 \leq i \leq n\}$.

$||\cdot||$ is the *standard norm* on k^n: for non-archimedean k, for $x = \{x_i\}_{1 \leq i \leq n} \in k^n$, $||x|| = \max\{|x_i|_k \mid 1 \leq i \leq n\}$;

when k is archimedean, $k \simeq \mathbb{R}$ *or* $\mathbb{C}$ and the standard norm $|| \cdot ||$ is defined by setting for $x = \{x_i\}_{1 \leq i \leq n} \in k^n$, $||x||_k = (\sum_{i=1}^{n} |x_i|_k^2)^{1/2}$.

For $x = \{x_i\}_{1 \leq i \leq n}$ and $y = \{y_i\}_{1 \leq i \leq n}$ in k^n, $d_k(x, y) = ||x - y||_k$.

If $T \in GL(n, \mathfrak{o}_k)$, $||T(v)|| = ||v||$ for all $v \in k^n$ (exercise).

If V is a vector space, an ordered basis $B = \{b_i \mid 1 \leq i \leq n\}$ gives an isomorphism $f_B : V \to k^n$; for $v \in V$, $||v||_B$ is defined as $||f_B(v)||$.

If B' is a basis such that $f_{B'} = f_B \circ T$ with $T \in GL(n, \mathfrak{o}_k)$, $||v||_{B'} = ||v||_B$ for all $v \in V$.

For $x = \{x_i\}_{1 \leq i \leq n}$ and $y = \{y_i\}_{1 \leq i \leq n}$ in k^n, $d_B(x, y) = ||x - y||_B$. Hence, when k is non-archimedean with residue field of characteristic p, $d_B(x, y)$ is an integral power of p^{1/e_k}.

1.92. Notation (Continued). For $x \in k^n$ and $d > 0$ in $\mathbb{R}$, $B_d(x) = \{y \in k^n \mid ||y - x|| < d\}$ is the *open disc of radius d around x* and

$\bar{B}_d(x) = \{y \in k^n \mid ||y - x|| \leq d\}$ is the *closed disc of radius d around x*.

When k is $\mathbb{R}$ or $\mathbb{C}$, $B_d(x)$ is the interior of $\bar{B}_d(x)$.

When k is non-archimedean $B_d(x), d > 0$ is open and compact in k^n (in fact given $d > 0$, there is a $d' \geq d$ such that $\bar{B}_d(x) = B_{d'}(x)$) – d' may be taken as the smallest integral power of p^{1/e_k} greater than d.

For such a d', there is an element $\lambda \in k$ such that $|l|_k = d'$. Further, $B_{d'}(0) = \lambda(\sum_{1 \leq i \leq n}(\mathfrak{o}_k \cdot e_i))$ so that $B - d(0)$ is a $\mathfrak{o}_k$ submodule of V.

If k is not archimedean and $y \in B_d(x)$, $B_d(y) = B_d(x)$. When $n = 1$, for $x \in k(= k^1)$, $B_d(x)$ (resp. $\bar{B}_d(x)$) is also denoted $I_d(x)$ (resp. $\bar{I}_d(x)$) – when $k = \mathbb{R}$, $I_d(x)$ (resp. $\bar{I}_d(x)$) is the open (resp. closed) interval $(x - d, x + d)$ (resp. $[x - d, x + d]$) whence the notation.

An *open cube* $Q_d(x)$ *of side d around* $x = \{x_i\}_{1 \leq i \leq n}$ in k^n is the set $\{y = \{y_i\}_{1 \leq i \leq n} \in k^n \mid y_i \in I_{d/2}(x_i)\}$.

1.93. For a non-archimedean k, $\mathfrak{o}_k$ is a principal ideal domain. We summarize in the theorem below a number of results on the structure of modules over $\mathfrak{o}_k$ which we will have occasion to use. The multiplication by elements of $\mathfrak{o}_k$ makes $\mathfrak{o}_k^r$ into a $\mathfrak{o}_k$-module.

Let M be a module over $\mathfrak{o}_k$. $m \in M$ is a torsion element if there is a $r \in \mathbb{N}$, $r > 0$ such that $\mathfrak{p}_k^r \cdot v = \{0\}$. The set $t(M)$ of torsion elements in M is the *torsion submodule* of M. M is *torsion-free* if $t(M) = \{0\}$. For an integer $r > 0$, k^r is a torsion-free $\mathfrak{o}_k$-module. A module M is a *free* module if $M \simeq \mathfrak{o}_k^q$ for some q.

1.94. Theorem. *Let $M \neq \{0\}$ be a finitely generated $\mathfrak{o}_k$-module.*
(i) *Then there are integers $q, r \in \mathbb{N}$ with $(q + r) > 0$ and positive integers $\{l_i \mid 1 \leq i \leq r\}$ such that M is the direct sum of q copies of $\mathfrak{o}_k$ and the mono-gene modules $\mathfrak{o}_k / \mathfrak{p}_k^{l_i}$, $1 \leq i \leq r$.*
(ii) *$t(M) \simeq \coprod_{i=1}^r \mathfrak{o}_k / \mathfrak{p}^{l_i}$.*
(iii) *M is torsion-free if and only if it is free and admits an injective $\mathfrak{o}_k$-module morphism into k^q for some $q \in \mathbb{N}$.*
(iv) *A submodule of a free module is free.*
(v) *Let V be a vector space over k. Let $B = \{e_i \mid 1 \leq i \leq r\}$ be a basis of V over k and $M = \sum_{i=1}^r \mathfrak{o}_k \cdot e_i$. Then $M \simeq \mathfrak{o}_k^r$. Let W be a vector subspace of V of dimension d. Then there is a basis $\{f_j \mid 1 \leq j \leq r\}$ such that $M = \sum_{i=1}^r \mathfrak{o}_k \cdot e_i$ and $W \cap M = \sum_{i=1}^d \mathfrak{o}_k \cdot e_i$.*

We will have occasions to use the following result (see Serre [14]).

1.95. Theorem. *Let k be a local field and $q > 0$ an integer. Then there is a finite extension K of k such that every extension of k of degree $\leq q$ is isomorphic over k to a subfield of K (containing k).*

1.96. Let V be a vector space of dimension q over the local field k. Then if B, B' are bases of V, there is a constant $C > 1$ such that $C^{-1} \cdot ||v||_B \leq ||v||_{B'} \leq C \cdot ||v||_B$. Let $T \in \mathrm{End}_k(V)$ and $\sigma(T) = \max\{|\lambda|_K \mid \lambda \text{ an eigen-value of } T\}$. Note by the theorem above, there is a finite extension K of k containing all the eigen-values of every $T \in \mathrm{End}_k(V)$. $\sigma(T)$ is the *spectral radius* of T. Let B be an ordered basis of V. For $T \in \mathrm{End}_k(V)$, set $||T||_B = \sup\{||T(v)||_B / ||v||_B \mid v \neq 0\}$. The following result is well known (and not difficult to prove).

1.97. Theorem. *Let V be a vector space and B an ordered basis. Let $T \in \mathrm{End}_k(V)$. Then the sequence $\{(||T^r||_B)^{1/r}\}_{r \in \mathbb{N}}$ converges to $\sigma(T)$.*

2. Analytic Functions

In this chapter we introduce the definition of a k-valued analytic function on an open set Ω in k^n where k is a local field k and prove the main results about them. When $k \simeq \mathbb{C}$ this is a familiar concept defined via the existence of the first derivative with respect to a complex variable. When $k = \mathbb{R}$ one may define a $\mathbb{R}$-valued analytic function on Ω as the restriction to Ω of a $\mathbb{C}$-valued analytic function on an open set $\tilde{\Omega}$ in $\mathbb{C}^n$ containing Ω. An alternative approach (following Weierstrass) is via convergent power series over $\mathbb{R}$. When k is not archimedean, only the second approach is available and is adopted in this chapter.

2.1. Formal Power Series. Let k be a field of characteristic 0. A *formal power series f in n variables over k* is a function $f : \mathbb{N}^n \to k$. For $\alpha \in \mathbb{N}^n$, let $X^\alpha : \mathbb{N}^n \to k$ be the power series defined by setting for $\beta \in \mathbb{N}^n$, $X^\alpha(\beta) = \delta_{\alpha,\beta}$. Set $X^0 = 1$ and for $1 \le i \le n$, $X_i = X^{<i>}$ where $< i > \in \mathbb{N}^n$ is the element $\{\delta_{i,j}| 1 \le j \le n\}$.

We denote a formal power series f by the more suggestive notation $\sum_{\alpha \in \mathbb{N}^n} f(\alpha)X^\alpha$. The αth *coefficient of f* is $f(\alpha)$; it is also denoted f_α. f_0 is the *constant term* of f. The set of all formal power series in n variables over k is denoted $k[[X]]$ with X denoting the n-tuple $\{X_i| 1 \le i \le n\}$. The $\{X_i| 1 \le i \le n\}$ are the *variables*.

The set $k[[X]]$ has a natural structure of a commutative k-algebra under the following operations: for $f,\ g\ \in k[[X]]$ and $\lambda \in k$ we define:

$$\sum_{\alpha \in \mathbb{N}^n} f_\alpha \cdot X^\alpha + \sum_{\alpha \in \mathbb{N}^n} g_\alpha \cdot X^\alpha = \sum_{\alpha \in \mathbb{N}^n} (f_\alpha + g_\alpha) \cdot X^\alpha$$

$$\left(\sum_{\alpha \in \mathbb{N}^n} f_\alpha \cdot X^\alpha\right) \cdot \left(\sum_{\alpha \in \mathbb{N}^n} g_\alpha \cdot X^\alpha\right) = \sum_{\alpha \in \mathbb{N}^n} \left(\sum_{\beta,\gamma \in \mathbb{N}^n, \beta+\gamma=\alpha} f_\beta \cdot g_\gamma\right) \cdot X^\alpha$$

$$\lambda \cdot \left(\sum_{\alpha \in \mathbb{N}^n} f_\alpha \cdot X^\alpha\right) = \sum_{\alpha \in \mathbb{N}^n} (\lambda \cdot f_\alpha) \cdot X^\alpha$$

It is easily checked that we have for $\alpha \in \mathbb{N}^n, X^\alpha = \prod_{1 \le i \le n} X_i^{\alpha_i}$. Also the map $\lambda \rightsquigarrow \lambda \cdot X^0$ of k in $k[[X]]$ imbeds k as a subring of $k[[X]](= k[[X_1, X_2, \cdots, X_n]])$ identifying 1 with X^0. We denote by $k[X]$ the k-subalgebra $\{f \in k$

$[[X]]|\{\alpha \in \mathbb{N}^n | f_\alpha \neq 0\}$ *is finite*}. The notation is justified as this subalgebra of $k[[X]]$ is isomorphic to the polynomial algebra in n variables over k with the $\{X_i | 1 \leq i \leq n\}$ playing the role of the n variables. In the sequel the polynomial algebra will be treated as a subalgebra of the power series ring $k[[X]]$.

2.2. Let $\epsilon : k[[X]] \to k$ be the *augmentation map* defined by $k[[X]] \ni f \rightsquigarrow f_0$. It is a k-algebra homomorphism of $k[[X]]$ *onto* k. It follows that kernel$(\epsilon) = \mathfrak{m}$ is a maximal ideal in $k[[X]]$. Evidently $k[[X]] = k \oplus \mathfrak{m}$. The ideal $\mathfrak{m}$ is easily seen to be generated by the $\{X_i | 1 \leq i \leq n\}$ and consequently, the ideal $\mathfrak{m}^2$ is generated by $\{X_i \cdot X_j | 1 \leq i \ j \leq n\}$.

It follows that $T^* = \mathfrak{m}/\mathfrak{m}^2$ is a k-vector space with the images $\{\bar{X}_i | 1 \leq i \leq n\}$ of $\{X_i | 1 \leq i \leq n\}$ in $\mathfrak{m}/\mathfrak{m}^2$ forming a basis. Thus $\dim_k T^* = n = \dim_k T (= \text{Hom}_k(T^*, k))$.

Let $v \in T$; then v defines a k-linear map (which we continue to denote v) of $\mathfrak{m}$ into k such that $v(\mathfrak{m}^2) = 0$. We extend $v : \mathfrak{m} \to k$ to a linear map $\tilde{v} : k[[X]](= k \oplus \mathfrak{m}) \to k$ by setting $\tilde{v} = 0$ on k and $\tilde{v}|_{\mathfrak{m}} = v$. The proof of the following lemma is left to the reader.

2.3. Lemma. *The map $T \ni v \rightsquigarrow \tilde{v}$ is a k-linear isomorphism of T on the space of all linear forms $L : k[[X]] \to k$ satisfying the following condition: for $f, f' \in k[[X]]$,*

$$L(f \cdot f') = L(f) \cdot \epsilon(f') + \epsilon(f) \cdot L(f'). \qquad (*)$$

2.4. Substitution. Suppose now that u is a formal power series in m variables and $F = \{f_i | 1 \leq i \leq m\}$ are m formal power series in n variables all belonging to $\mathfrak{m}$. The formal power series $u(F) = u(f_1, f_2, \cdots, f_n)$ in n variables *obtained by substituting the $\{f_i | 1 \leq i \leq m\}$ in u* is defined as follows: for $\alpha \in \mathbb{N}^n$,

$$u(f_1, f_2, \cdots, f_m)_\alpha = \sum_{\beta \in \mathbb{N}^m} u_\beta \cdot \Big(\prod_{1 \leq i \leq m} f_i^{\beta_i} \Big)_\alpha.$$

Note that the summation on the right hand side is over only finitely many terms:
$(\prod_{1 \leq i \leq m} f_i^{\beta_i})_\alpha = 0$ if $|\beta| > |\alpha|$ since $f_{i0} = 0$ for $1 \leq i \leq m$.

2.5. Definitions. The *exponential* series $\exp(X)$ is the power series (in one variable) $\sum_{n=0}^{\infty} X^n/n!$. The *logarithmic* series $\ln(1 + X)$ is the power series (in one variable) $\sum_{n=1}^{\infty} (-1)^{n-1} \cdot X^n/n$. For $e \neq 0$ in k, $(1+X)^e$, the *binomial series of exponent e*, is the formal power series $\sum_{r=0}^{\infty} \binom{e}{r} \cdot X^r$ (see 1.2 for notation). We also set $(1 + X)^0 = 1$.

2.6. Lemma. (i) *If $f, g \in \mathfrak{m}$, $\exp(f) \cdot \exp(g) = \exp(f + g)$.*
(ii) *For $e, f \in k$, $(1 + X)^e \cdot (1 + X)^f = (1 + X)^{(e+f)}$.*

2.7. Proof. (i) is left as an exercise. To prove (ii) we need to show that for every $r \in \mathbb{N}$

$$\binom{(e+f)}{r} = \sum_{p,q \in \mathbb{N}, (p+q)=r} \binom{e}{p} \cdot \binom{f}{q}. \tag{*}$$

This is a well-known fact when e and f are positive integers. Fix $r > 0$ in $\mathbb{N}$. Then for a fixed $e \in \mathbb{Z}$,

$$P_{e,r}(f) = \binom{(e+f)}{r} - \sum_{p,q \in \mathbb{N}, (p+q)=r} \binom{e}{p} \cdot \binom{f}{q}$$

is a polynomial function of f which vanishes on all of $f \in \mathbb{Z}^+$ and hence is identically zero. Thus $P_{e,r}(f) = 0$ for any $f \in k$ and $e \in \mathbb{Z}^+$. Now for fixed f, $P_{e,r}(f)$ is a polynomial function of e which is zero for all $e \in \mathbb{Z}^+$ and is therefore zero for all $e \in k$. Hence (iii) of the lemma.

2.8. If $f \in k[[X]]$ (n variables) with $f_0 = 0$, we denote the power series obtained by substituting f in $(1+X)^e$, $(1+f)^e$. In particular if we substitute the power series $f \in \mathfrak{m} \subset k[[Y]]$ (in any number of variables Y) in the power series $\sum_{0 \le r < \infty} X^r$ in one variable, we get a power series which is easily seen to be the the the inverse of $(1-f)$: $(1-f)(\sum_{0 \le r < \infty} f^r) = 1$. This evidently implies the following.

2.9. Lemma. $f \in k[[X]]$ *is invertible and only if* $f_0 \ne 0$. $k[[X]]$ *is a 'local ring' with* $\mathfrak{m}$ *as its unique maximal ideal.*

2.10. Next, we define for each i with $1 \le i \le n$, a vector space endomorphism $D_i : k[[X]] \to k[[X]]$ as follows: for $f \in k[[X]]$,

$$D_i(f)_\alpha = (\alpha_i + 1) \cdot f_{(\alpha + <i>)}.$$

D_i also denoted D_{X_i}. It is easy to see that one has for f, g in $k[[X]]$,

$$D_i(f \cdot g) = D_i(f) \cdot g + f \cdot D_i(g).$$

Observe that the linear forms $\epsilon \circ D_i$, $1 \le i \le n$, satisfy $(*)$ of 2.3 and form the dual basis to $\{\bar{X}_i | 1 \le i \le n\}$ (see 2.2). One sees from the definitions that for $1 \le i, j \le n$, $D_i \circ D_j = D_j \circ D_i$. More generally we define for $\alpha \in \mathbb{N}^n$, the operator D^α as the endomorphism $D_1^{\alpha_1} \circ D_2^{\alpha_2} \cdots \circ D_n^{\alpha_n}$ of $k[[X]]$, D_i^q being the endomorphism of $k[[X]]$ obtained by q times repeated composition of D_i with itself. In the sequel we will denote $D^\alpha \circ D^\beta$ by $D^\alpha \cdot D^\beta$. The operators $\{D^\alpha | \alpha \in \mathbb{N}^n\}$ are easily seen to commute with one another and one has $D^\alpha \cdot D^\beta = D^{(\alpha+\beta)}$ for all $\alpha, \beta \in \mathbb{N}^n$. For $\alpha \in \mathbb{N}^n$, the series $D^\alpha(f)$ is the αth *derivative* of f. One checks easily (by induction on $|\alpha|$) that for $f \in k[[X]]$, $f_\alpha = (D^\alpha(f))_0 / \alpha!$.

Let k' be a field containing k. The inclusion $h : k \to k'$ induces an injective k-algebra homomorphism $\tilde{h} : k[[X]] \to k'[[X]]$:

$$\tilde{h}\left(\sum_{\alpha \in \mathbb{N}^n} f_\alpha \cdot X^\alpha \right) = \sum_{\alpha \in \mathbb{N}^n} h(f_\alpha) \cdot X^\alpha.$$

Then $\tilde{h}(\{f \in k[[X]] \mid f_0 = 0\}) \subset \{f \in k'[[X]] \mid f_0 = 0\}$. Also, if u is a formal power series in m variables over k and $\{f_i \mid 1 \le i \le m\}$ are m formal power series in n variables over k with $f_{i0} = 0$, then $\tilde{h}(u(f_1, f_2, \cdots, f_m)) = \tilde{h}(u)(\tilde{h}(f_1), \tilde{h}(f_2), \cdots \tilde{h}(f_m))$. For $f \in k[[X]]$, and $1 \le i \le n$, $D_i(\tilde{h}(f)) = \tilde{h}(D_i(f))$.

2.11. Lemma. *Let φ be a power series in n variables $\{Y_i\}_{1 \le i \le n}$ and $F = \{f_i \mid 1 \le i \le n\}$ be n power series in $k[[X = \{X_j \mid 1 \le j \le m\}]]$ all belonging to the maximal ideal $\mathfrak{m}$ of $k[[X]]$. Then for $1 \le j \le m$,*

$$D_{X_j}(\varphi(F)) = \sum_{i=1}^{n}(D_{Y_i}(\varphi)(F)) \cdot D_{X_j} f_i$$

The proof is easily reduced to the case when φ is a monomial Y^α for $\alpha \in \mathbb{N}^m$; and in this special case it is left as an exercise.

2.12. Corollary. *For power series $f \in k[[x]]$ in one variable X, we have:*
(i) $D_X(\exp(X)) = \exp(X)$.
(ii) $D_X(\ln(1 + X)) = (1 + X)^{-1} (= \sum_{n=0}^{\infty}(-1)^n \cdot X^n)$.
(iii) $\ln(1 + \exp(X) - 1) = X$.
(iv) $\exp(\ln(1 + X)) = 1 + X$.

2.13. Proof. (i) and (ii) are obvious from the definitions. Set $u = \ln(1 + (\exp - 1)) = \sum_{n=0}^{\infty} u_n \cdot X^n$. It is easy to see that $a_o = 1$. So to prove (iii), we need to show that $u_n = 0$ for all $n \in \mathbb{N}$, $n > 0$. Now by the above lemma combined with (i) and (ii) we have:

$$\sum_{n \in \mathbb{N}}(n + 1) \cdot u_n X^n = \exp(-X) \cdot \exp(X) = 1$$

from which it follows that $u_n = 0$ for all $n \in \mathbb{N}$, $n > 0$. Set $v = \exp(\ln(1 + x)) = \sum_{n=0}^{\infty} v_n \cdot X^n$; from (i) and (ii) above and 2.11,

$$\sum_{n=0}^{\infty}(n + 1)v_{n+1} \cdot X^n = X \cdot v = \exp(\ln(1 + X)) \cdot (1 + X)^{-1}$$

so that $(1 + X)(\sum_{n=0}^{\infty}(n + 1)v_{n+1} \cdot X^n) = \sum_{n=0}^{\infty} v_n \cdot X^n$; it follows that $v_0 = v_1 = 1$ and $v_n = 0$ for $n \ge 2$.

In the rest of this chapter and the next two chapters, k will be a *local field* with its (standard) absolute value $| \cdot |_k$ (see 1.85−1.93 for notation).

2.14. Proposition. *The following four properties of a formal power series f over k are equivalent:*

(i) $\exists\ \delta > 0$ *such that the (real) series* $\sum_{\alpha \in \mathbb{N}^n} |f_\alpha|_k \cdot \delta^{|\alpha|}$ *converges.*

(ii) $\exists\ x \in k^{\times n}$ *such that* $\sum_{\alpha \in \mathbb{N}^n} f_\alpha \cdot x^\alpha$ *converges to a limit (in k). In other words* $\exists\ y \in k$ *such that* $\forall \epsilon > 0,\ \exists N \in \mathbb{N}$ *such that* $|y - \sum_{\alpha \in \mathbb{N}, |\alpha| \leq r} f_\alpha \cdot x^\alpha| \leq \epsilon$ *for all* $r \geq N$.

(iii) $\exists\ \delta > 0$ *such that* $|f_\alpha|_k \cdot \delta^{|\alpha|}$ *tends to 0 (in $\mathbb{R}$) as $|\alpha|$ tends to ∞.*

(iv) $\exists\ C > 0$ *and* $\delta > 0$ *such that* $|f_\alpha|_k \leq C \cdot \delta^{-|\alpha|}$ *for all* $\alpha \in \mathbb{N}^n$.

2.15. Proof. That (i)$\Rightarrow$(ii)$\Rightarrow$(iii)$\Rightarrow$(iv) is obvious. If (iv) holds, then $\sum_{\alpha \in \mathbb{N}^n} |f_\alpha|_k \cdot \eta^{|\alpha|} < \infty$ if $\eta > 0$ is such that $\delta^{-1} \cdot \eta < 1$, hence (i).

2.16. Definitions. A formal power series $f \in k[[X]]$ is *convergent* if it satisfies one of the (equivalent) conditions (i)–(iv) of Proposition 2.14. The *radius of convergence* $\rho(f)$ of a convergent power series f is defined to be $\sup\{\rho > 0|\ |f_\alpha| \cdot \rho^{|\alpha|}$ *tends to 0 as $|\alpha|$ tends to ∞*$\}$.

2.17. Proposition. *The set $k\{\{X\}\}$ of convergent power series is a subring of $k[[X]]$: For $f, f' \in k\{\{X\}\}$, $\rho(f + f')$ and $\rho(f \cdot f')$ are $\geq \inf(\rho(f), \rho(f'))$. If $f \in k[X]$, $\rho(f) = \infty$ so that $k[X] \subset k\{\{X\}\}$.*

2.18. Proof. f and f' being convergent, there exists $C, \delta > 0$ such that $|f_\alpha|_k, |f'_\alpha|_k \leq C \cdot \delta^{-|\alpha|}$. Let $F \in \mathbb{R}[[X]]$ be the power series $\sum_{\alpha \in \mathbb{N}^n} C \cdot \delta^{|\alpha|} \cdot X^\alpha$. Clearly $F \in \mathbb{R}\{\{X\}\}$ and $\rho(F) = \delta^{-1}$; also $|f_\alpha|_k, |f'_\alpha|_k \leq F_\alpha$ and hence $|(f + f')_\alpha|_k \leq 2 \cdot F_\alpha$ and $|(f \cdot f')_\alpha|_k \leq (F^2)_\alpha$. Thus we only need to show that $\rho(2 \cdot F), \rho(F^2) \geq \rho(F)$. This is obvious in the case of $2 \cdot F$. By definition $(F^2)_\alpha = C^2 \cdot (\sum_{\beta, \gamma \in \mathbb{N}^n, \beta + \gamma = \alpha} \delta^{|\beta + \gamma|}) = C^2 \cdot (\prod_{1 \leq i \leq n} (\alpha_i + 1)) \cdot \delta^{|\alpha|} \leq C^2 \cdot (|\alpha| + 1)^n \cdot \delta^{|\alpha|}$. Now $A^{-|\alpha|} \cdot (|\alpha| + 1)^n \to 0$ as $|\alpha| \to \infty$ for any $A > 1$. Hence $\exists\ c(A) > 0$ such that $(|\alpha| + 1) \leq c(A) \cdot A^{|\alpha|}$. Thus we find that $(F^2)_\alpha \leq C^2 \cdot c(A) \cdot A^{|\alpha|} \cdot \delta^{|\alpha|}$ so that $\rho(F^2) \geq A^{-1} \cdot \rho(F)$. As this holds for every $A > 1$, $\rho(F^2) \geq \rho(F)$. That $\rho(f) = \infty$ for $f \in k[X]$ is obvious.

2.19. Proposition. *If $f \in k\{\{X\}\}$, then for $\alpha \in \mathbb{N}^n$, $D^\alpha f \in k\{\{X\}\}$ and $\rho(D^\alpha f) \geq \rho(f)$. In particular the k-subalgebra $k\{\{X\}\}$ of $k[[X]]$ is stable under the endomorphism D^α.*

2.20. Proof. Induction on $|\alpha|$ reduces the proof to the case $|\alpha| = 1$. We need to show that $\rho(D_i(f)) \geq \rho(f)$ for $1 \leq i \leq n$. This follows from the fact that $|q|_k \leq q$ for all $q \in \mathbb{N}$, and for $\mathbb{R} \ni A > 1$, there is a $b > 0$ with $q \leq b \cdot A^q$ for all $q \in \mathbb{N}$.

2.21. Proposition. *Let $u \in k\{\{Y_1, Y_2, \cdots, Y_m\}\}$ and $\{f_i|\ 1 \leq i \leq m\} \subset k\{\{X_1, X_2, \cdots, X_n\}\} \cap \mathfrak{m}$. Then the formal power series $u(f_1, f_2, \cdots, f_m)$ (cf. 2.4) is convergent.*

2.22. Proof. We may assume, without that $|u_0|_k \leq 1$. As the $\{f_i | 1 \leq i \leq m\}$ and u are convergent, $\exists \; \delta > 0$ (in $\mathbb{R}$) such that $|u_\beta|_k \leq \delta^{|\beta|}$, $\forall \beta \in \mathbb{N}^m$ and $|f_{i\beta}|_k \leq \delta^{|\beta|}$ for $1 \leq i \leq m$ and $\forall \beta \in \mathbb{N}^m$: note that $|u_0|_k \leq 1$ and $|f_{i0}| = 0$ for $1 \leq i \leq m$. Let U (resp. F) be the power series $\sum_{\beta \in \mathbb{N}^m} \delta^{|\beta|} \cdot Y^\beta$ (resp. $\sum_{\alpha \in \mathbb{N}^n} \delta^{|\alpha|} \cdot X^\alpha$) in m (resp. n) variables over $\mathbb{R}$. Let $F_i = F$ for $1 \leq i \leq m$. Then $U \in \mathbb{R}\{\{Y\}\}$ and $F_i \in \mathbb{R}\{\{X\}\}$ for $1 \leq i \leq m$. One then has:

$$|u(f_1, f_2, \cdots, f_m)_\alpha|_k \leq U(F_1, F_2, \cdots, F_m)_\alpha$$

for all $\alpha \in \mathbb{N}^n$. Thus it suffices to show that $U(F_1, F_2, \cdots, F_m)$ is convergent. To see this let $t > 0$ be such that $\delta \cdot t = \eta < 1$. Then for $1 \leq i \leq m$, the infinite series (of *positive* terms) $\sum_{\alpha \in \mathbb{N}^n} F_{i\alpha} \cdot t^{|\alpha|}$ converges to $\zeta = (1 - \eta)^{-n}$. Choose t small enough to ensure that $\zeta \leq \delta^{-1}$. Then the multiple infinite series of positive terms (obtained by expanding the terms under the product signs) in $\sum_{\beta \in \mathbb{N}^m} U_\beta \cdot \prod_{1 \leq i \leq m} (\sum_{\alpha \in \mathbb{N}^n} F_{i\alpha} \cdot t^{|\alpha|})^{\beta_i}$ converges. Now $\sum_{\alpha \in \mathbb{N}^n} U(F_1, F_2, \cdots, F_m)_\alpha t^{|\alpha|}$ is the result of a rearrangement of the same infinite series of positive real numbers followed by adding up suitable groups of terms and is thus finite proving that $U(F_1, F_2, \cdots, F_m)$ is a convergent power series.

2.23. Corollary. *A power series $f \in k\{\{X\}\}$ is invertible in $k\{\{X\}\}$ if and only if $f_0 \neq 0$, i.e., $f \notin \mathfrak{m}$. (Consequently) $k\{\{X\}\}$ is a local k-algebra with $\mathfrak{m}_o = \mathfrak{m} \cap k\{\{X\}\}$ as its (unique) maximal ideal.*

2.24. Proof. If f is invertible and g its inverse, then $1 = (g \cdot f)_0 = g_0 \cdot f_0$, hence $f_0 \neq 0$. Conversely assume that $f_0 \neq 0$. It suffices to show that $f_0^{-1} \cdot f$ is invertible in $k\{\{X\}\}$: in other words we may assume that $f_0 = 1$. Now $1 - f$ is a convergent power series and is in $\mathfrak{m}$. If we substitute it in the convergent power series $u = \sum_{n \in \mathbb{N}} X^n$, one gets a convergent power series which is the inverse of f in $k\{\{X\}\}$ – as it is the inverse of f in $k[[X]]$ (see 2.8–2.9).

2.25. The $\{X_i | 1 \leq i \leq n\}$ which generate $\mathfrak{m}$ as an ideal in $k[[X]]$ belong to $k\{\{X\}\}$ and hence to $\mathfrak{m}_o$. It is an easy exercise to deduce that they generate $\mathfrak{m}_o$ as an ideal in $k\{\{X\}\}$. Further $\mathfrak{m}_o^2 = \mathfrak{m}^2 \cap k\{\{X\}\}$ is generated by $\{X_i \cdot X_j | 1 \leq i, j \leq n\}$. It follows that the natural map $\mathfrak{m}_o/\mathfrak{m}_o^2 \to \mathfrak{m}/\mathfrak{m}^2$ $(= T^*)$ is an isomorphism (see 2.2–2.3). Moreover the dual T of T^* can be identified with the set of all linear maps $L : k\{\{X\}\} \to k$ satisfying the following condition: for f, f' in $k\{\{X\}\}$,

$$L(f \cdot f') = L(f) \cdot f_0' + f_0 \cdot L(f').$$

2.26. Recall that for $x = \{x_i\}_{1 \leq i \leq n} \in k^n$, $||x||_k = \max\{|x_i|_k \mid 1 \leq i \leq n\}$ if k is non-archimedean while, if k is archimedean, $||x||_k = (\sum_{i=1}^n |x_i|^2)^{1/2}$. For $x \in k^n$ and $\delta > 0$, $B(x, \delta)$ or $B_\delta(x)$ (resp. $\bar{B}(x, \delta)$ or $\bar{B}_\delta(x)$) denotes the open (resp. closed) disc $\{y \in k^n | \; ||y - x||_k < \delta\}$ (resp. $\{y \in k^n | \; ||y - x||_k \leq \delta\}$)

of radius δ around x . Suppose now that f is a convergent power series with radius of convergence $\rho(=\rho(f))$ and $x_o \in k^n$, then the series $\sum_{\alpha \in \mathbb{N}^n} f_\alpha \cdot (x - x_o)^\alpha$ converges uniformly and absolutely in $\bar{B}_\delta(x_o)$ for $\delta < \rho$ and hence defines in the open disc $B_\rho(x_o)$ a continuous (k-valued) function which we will denote $_{x_o} f$.

2.27. Recall the definition of the *germ* of a k-valued function at a point p in a topological space X: let $\mathcal{P}$ be the set of all pairs (U, h) where U is an open neighbourhood of p in X and h is a k-valued function on U; define the equivalence relation $\sim$ on $\mathcal{P}$ by setting for (U, h), (U', h') in $\mathcal{P}$, $(U, h) \sim (U', h')$ iff $h = h'$ on some open neighbourhood V of p (contained in $U \cap U'$). An equivalence class in $\mathcal{P}$ is a *germ of a function* at p. The germ determined by $(U, h) \in \mathcal{P}$ is the *germ of h*. We denote by $\dot{h}_p$ the germ determined by the pair (U, h) at p and by $\mathcal{F}_p$ the set of all germs of k-valued functions at p. One has a structure of a k-algebra on $\mathcal{F}_p$: for $\lambda \in k$ and pairs (U, h) and (U', h') in $\mathcal{P}$, $\lambda \cdot \dot{h}_p$ (resp. $\dot{h}_p + \dot{h}'_p$, resp. $\dot{h}_p \cdot \dot{h}'_p$) is the equivalence class of $(U, \lambda \cdot h)$ (resp. $(U \cap U', h + h')$, resp. $(U \cap U', h \cdot h')$). It is clear that if $(U, h) \sim (U', h')$, then $h(p) = h'(p)$ and thus the map $(f, h) \rightsquigarrow f(p)$ of $\mathcal{P}$ in k descends to a k-algebra homomorphism $ev_p : \mathcal{F}_p \to k$, the '*evaluation*' at p.

2.28. Let f be a convergent power series. Suppose that $x_o, y_o \in k^n$ are such that $||x_o - y_o||_k = \delta < \rho(f)$. Then one has for $y \in k^n$ with $||y - y_o||_k = \eta < (\rho(f) - \delta)$, $||y - x_o||_k \leq ||y - y_o||_k + ||y_o - x_o|_k < \eta + \delta$. It follows that we have $|(y - y_o)^\alpha|_k = |((y - y_o) + (y_o - x_o))^\alpha|_k = |\sum_{\beta \in \mathbb{N}^n, \beta \leq \alpha} \binom{\alpha}{\beta} \cdot (y - y_o)^\beta \cdot (y_o - x_0)^{(\alpha - \beta)}|_k \leq \sum_{\beta \in \mathbb{N}^n, \beta \leq \alpha} \binom{\alpha}{\beta} \eta^{|\beta|} \cdot \delta^{|(\alpha - \beta)|} = (\eta + \delta)^{|\alpha|}$ (note that for a non-negative integer q, $|q|_k \leq q$). It follows that the terms of the series $\sum_{\alpha \in \mathbb{N}^n} f_\alpha \cdot (\sum_{\beta \in \mathbb{N}^n, \beta \leq \alpha} \binom{\alpha}{\beta} \cdot (y - y_o)^\beta \cdot (y_o - x_0)^{(\alpha - \beta)})$ rearranged in any manner converge to $_{x_o} f(y)$ for all $y \in k^n$ with $|y - y_o|_k < (\rho(f) - \delta)$. Thus we have for $y \in k^n$ with $|y - y_o|_k < (\rho(f) - \delta)$,

$$_{x_o} f(y) = \sum_{\beta \in \mathbb{N}^n} \left\{ \sum_{\alpha \in \mathbb{N}^n, \alpha \geq \beta} f_\alpha \cdot \binom{\alpha}{\beta} (y_o - x_o)^{(\alpha - \beta)} \right\} \cdot (y - y_o)^\beta$$

$$= \sum_{\beta \in \mathbb{N}^n} g_\beta \cdot (y - y_o)^\beta$$

where we have set for $\beta \in \mathbb{N}^n$, $g_\beta = \sum_{\alpha \in \mathbb{N}^n, \alpha \geq \beta} \binom{\alpha}{\beta} (y_o - x_o)^{(\alpha - \beta)}$. Evidently $g = \sum_{\beta \in \mathbb{N}^n} g_\beta \cdot X^\beta$ is a convergent power series with radius of convergence $\rho(g) \geq \rho(f) - \delta$ and $_{x_o} f(y) = {_{y_o}} g(y)$ for all $y \in B_\delta(x_o)$ with $|y - y_o| < \rho(f) - \delta \ (\leq \rho(g))$.

2.29. Definition. A k-valued function f on an open set U in k^n is *analytic* in U if the following holds: for every $y \in U$ $f^y \in k\{\{X\}\}$ and there is an open set V with $y \in V \subset U$ such that for all $z \in V$, $f(z) = {_y}f^y(z)$. A function $F : U \to E$ where E is a finite dimensional k-vector space is *analytic*, if $L \circ F$

is analytic for every linear form $L : E \to k$ on E. Equivalently if we identify E with k^r through a basis of E over k and F with an r-tuple $\{F_i\}_{1 \leq i \leq r}$ of k-valued functions, F is *analytic* iff the $\{F_i\}_{1 \leq i \leq r}$ are analytic.

2.30. Remarks. (1) An analytic function is necessarily continuous. The discussion in 2.27 shows that for a convergent power series f and $x_o \in k^n$, $_{x_o}f$ is an analytic function in $B_{\rho(f)}(x_o)$. A polynomial function is analytic on all of k^n.

(2) The set of k-valued analytic functions on an open set U in k^n is a k-algebra. Let φ and φ' be k-valued analytic functions on the open set U. Then for any point $p \in U$ there are convergent power series f and f' and an open V set contained in U with $p \in V$ such that $_pf$ (resp. $_pf'$) is defined in V and $_pf = \varphi$ (resp. $_pf' = \varphi'$) on V. Clearly then one has $_p(f + f') = (\varphi + \varphi')$ and $_p(f \cdot f') = \varphi \cdot \varphi'$ on V.

(3) The germs of analytic functions at a point $p \in k^n$ constitute a subring of $\mathcal{F}_p$ denoted $\mathcal{A}_p$. It is the direct limit of the k-vector spaces $\mathcal{A}(U)$ of analytic functions on open neighbourhoods U of p and the restriction maps $\mathcal{A}(U) \to \mathcal{A}(V)$ for inclusions $V \subset U$. From the definition of analytic functions, one concludes that $\mathcal{A}_p$ *is isomorphic to the k-algebra $k\{\{X\}\}$ of convergent power series in n variables.*

(4) Let $F : U \to E$ be an analytic function on an open set U in k^n taking values in a finite dimensional vector space E. Let x_o be any point in k^n and set $U' = \{y \in k^n | \ y + x_o \in U\}$. Let $L_{x_o} : k^n \to k^n$ be the translation by x_o: for $z \in k^n$, $L_{x_o}(z) = z + x_o$. Then the function $F \circ L_{x_o}$ is analytic. This is immediate from the definitions and the fact that for z, z_o in U' one has $(z - z_o) = ((z + x_o) - (z_o + x_o)) = (y - y_o)$ where $y = (z + x_o)$ and $y_o = (z_o + x_o)$ are in U.

(5) Analyticity is a local property: a function $f : U \to k$ on an open set in k^n is analytic if and only if for every point $p \in U$ there is an open neighbourhood U_p such that $f|_{U_p}$ is analytic.

(6) Let $f : U \to k$ be an analytic function on an open set U in k^n. For any point $p \in U$ we have a convergent power series f^p in $k\{\{X\}\}$ and an open disc V_p around p contained in U such that the series $\sum_{\alpha \in \mathbb{N}^n} f^p_\alpha(x - p)^\alpha$ converges absolutely and uniformly to $f(x)$ for all $x \in V_p$. Suppose now that K is a locally compact field containing k; then as $k\{\{X\}\}$ is contained in $K\{\{X\}\}$, f^p may be regarded as an element of $K\{\{X\}\}$. Further since $k^n \subset K^n$ one has an open disc W_p around p in K^n such that the series $\sum_{\alpha \in \mathbb{N}^n} f^p_\alpha \cdot (x - p)^\alpha$ converges absolutely and uniformly for all $x \in W_p$. It is not difficult to see that we can find a countable set $\{p_n\}_{n \in \mathbb{N}}$ such that $U = \bigcup_{n \in \mathbb{N}} V_{p_n}$ and $W_{p_n} \supset V_{p_n}$

such that $_{p_n} f^{p_n} =_{p_m} f^{p_m}$ on $W_m \cap W_n$ for every pair $(m,n) \in \mathbb{N}^n \times \mathbb{N}^n$. Thus if we set $\tilde{U} = \bigcup_{n \in \mathbb{N}} W_{p_n}$, the $_{p_n} f^{p_n}$ 'patch' together to define a K-valued analytic function $\tilde{f}$ on $\tilde{U}$ which restricts to f on U. It is not difficult to see that $\tilde{f}$ and $\tilde{f}'$ are analytic functions on open sets $\tilde{U}$ and $\tilde{U}'$ in K^n respectively both containing U and $\tilde{f} = \tilde{f}' = f$ on U, then there is an open set $\tilde{U}^* \subset \tilde{U} \cap \tilde{U}'$ open in K^n on which $\tilde{f} = \tilde{f}'$.

$(\mathbf{7})$ Let $\hat{\mathbb{Q}}$ be the closure of $\mathbb{Q}$ in k and k' a subfield of k containing $\hat{\mathbb{Q}}$. Let U be an open set in k^n and $f : U \to k$ be an analytic function. Now k may be viewed as a vector space over k'. Thus f may be considered a k'-vector space valued function on U (now treated as an open set in the $(r \cdot n)$-dimensional k'-vector space , where $r = \dim_{k'} k$). From the definitions it is easily seen that f as a k'-vector space valued function is analytic over k'. We thus have a k'-linear map $R_{k/k'}$ of the space of k-valued analytic functions (over k) on U into the k'-vector space valued analytic functions (over k') on U; in other words, we have a k'-linear map of $\mathcal{A}_p \simeq k\{\{X\}\}$ of germs of k-valued analytic functions (over k) at p into the space of germs of $(k'$-vector space $k)$-valued functions analytic (over k'); and this has an identification with $k \otimes_{k'} k'\{\{Y\}\}$ where Y represents $r \cdot n$ variables.

$(\mathbf{8})$ We continue with the notation in (7). Suppose now $L : k'\{\{Y\}\} \to k'$ is a linear map satisfying $L(f \cdot f') = L(f) \cdot \epsilon(f') + \epsilon(f) \cdot L(f')$; then L defines a k-linear map $1 \otimes L : k \otimes_{k'} k'\{\{Y\}\} \to k \otimes_{k'} k'$. The map $L \rightsquigarrow (1 \otimes L)$ composed with $R_{k/k'}$ defines an isomorphism of the k'-dual $T_{k'\{\{Y\}\}}$ of $\mathfrak{m}_{o_{k'}\{\{Y\}\}}/\mathfrak{m}^2_{o_{k'}\{\{Y\}\}}$ on the k-dual $T_{k\{\{Y\}\}}$ of $\mathfrak{m}_{o_k\{\{Y\}\}}/\mathfrak{m}^2_{o_k\{\{Y\}\}}$. We omit the proof which is straightforward.

$(\mathbf{9})$ Suppose that f is an analytic function on an open set Ω in k^n. Then for $x \in \Omega$, there is a $f' \in k\{\{X\}\}$ such that $f = _x f'$ in an open neighbourhood) U of $x \in \Omega$. We know that the function $_x f'$ is defined on all of $B_{\rho(f')}(x)$ and so we have necessarily $U \subset B_{\rho(f')}(x)$. If k is non-archimedean, for an $x \in \Omega$, $_x f'$ need not equal f on *all* of $B_{r(x)}(x) \cap \Omega$. The following example illustrates this phenomenon: let h be a convergent power series with radius of convergence ρ. Define f on $B_\rho(0)$ as follows: on $B_{\rho/2}(0)$, $f = _0 h$; on the complement of (the open and closed set) $B_{\rho/2}(0)$, $f = 0$.

2.31. Theorem. *Let* $F = \{f_i | 1 \leq i \leq m\} \subset \mathfrak{m}_o$ *be* m *(convergent) power series in* n *variables* X *and* Φ *a convergent power series in* m *variables* Y. *Then* $\Phi(F) = \Phi(f_1, f_2, \cdots, f_m)$ *is a convergent power series and there is a* $\delta \in \mathbb{R}^+$ *with* $\delta < \min(\{|\rho(f_i)|1 \leq i \leq m\}, \rho(\Phi(f)))$ *such that* $_0 F(B^n_\delta(0)) \subset B^m_{\rho(\Phi)}(0)$ *and* $_0(\Phi(F)) = (_0\Phi)_{\circ} F$ *on* $B^n_\delta(0)$.

2.32. Proof. We have already seen (2.20) that $\Phi(F) \in k\{\{X\}\}$. As the function $_0 F = \{_0 f_i\}_{1 \leq i \leq m}$ is continuous, there is a $\delta > 0$ with $\delta < \inf\{\rho(f_i) | 1 \leq i \leq m\}$ such that $_0 F(B^n_\delta(0)) \subset B^m_{\rho(\Phi)}(0)$. It follows that for $x \in B^n_\delta(0)$,

$$(_0\Phi)\circ(_0F)(x) = \sum_{\beta\in\mathbb{N}^m} \Phi_\beta \cdot (\prod_{j=1}^m {}_0f_i(x)^{\beta_i})$$

$$= \sum_{\beta\in\mathbb{N}^m} \Phi_\beta \cdot \{\sum_{\{\gamma(i)\in(\mathbb{N}^n)^{\beta_i},1\le i\le m\}} (\prod_{i=1}^m (\prod_{j=1}^{\beta_i} f_{i\gamma(i)_j})) \cdot x^{\sum_{i=1}^m (\sum_{j=0}^{\beta_i} \gamma(i)_j)}\}$$

Now the (multiple) infinite series of positive real numbers

$$\sum_{\beta\in\mathbb{N}^m} \Phi_\beta \cdot \{\sum_{\{\gamma(i)\in(\mathbb{N}^n)^{\beta_i},1\le i\le m\}} \prod_{i=1}^m (\prod_{j=1}^{\beta_i} |f_{i\gamma(i)_j}|_k) \cdot |x|_k^{\sum_{i=1}^m (|\sum_{j=0}^{\beta_i} \gamma(i)_j|)}\}$$

is convergent in view of our choice of δ. It follows that the series

$$\sum_{\beta\in\mathbb{N}^m} \Phi_\beta \cdot \{\sum_{\{\gamma(i)\in(\mathbb{N}^n)^{\beta_i},1\le i\le m\}} \prod_{i=1}^m (\prod_{j=1}^{\beta_i} f_{i\gamma(i)_j} \cdot x^{\sum_{i=0}^m (\sum_{j=0}^{\beta_i} \gamma(i)_j)})\}$$

rearranged in any manner converges to the limit $(_0\Phi)\circ(_0F)(x)$ for all $x \in B_\delta^n(0)$. But by the very definition of $\Phi(F)$, $\sum_{\alpha\in\mathbb{N}^n} \Phi(F)_\alpha \cdot x^\alpha$ being one such rearrangement followed by suitable grouping together of terms we conclude that $\sum_{\alpha\in\mathbb{N}^n} \Phi(F)_\alpha \cdot x^\alpha$ converges to $(_0\Phi)\circ(_0F)(x)$ for all $x \in B_\delta^n(0)$. This proves the theorem.

2.33. Theorem. *Let Ω, Ω' be open sets in k^p, k^q respectively. Let $F : \Omega \to \Omega'$ and $G : \Omega' \to k^r$ be analytic functions. Then $G\circ F$ is analytic.*

2.34. Proof. If $G = (g_1, g_2, \cdots, g_r)$ it suffices to show that $g_i\circ F$ is analytic for every i with $1 \le i \le r$. Thus we may assume that $r = 1$ and in this case the theorem is an immediate corollary of Theorem 2.30 and the fact that analyticity is a local property (see 2.30 (5)).

Our next result asserts that an analytic function on Ω (an open set in k^n) is 'differentiable' and all the 'partial derivatives' are analytic.

2.35. Proposition. *Let $f \in k\{\{X\}\}$ with radius of convergence ρ. Let $x_o \in k^n$ and $x \in B_\rho^n(x_o) = U$. Let $\{e_i\}_{1\le i\le n}$ be the standard basis of k^n. Then for $x \in U$, the $\lim_{h\to 0}(_{x_o}f(x + h \cdot e_i) - {}_{x_o}f(x))/h$ exists and equals $_{x_o}(D_if)(x)$ (cf. 2.10 and 2.18).*

2.36. Proof. Set $U = B_\delta(x_o)$ with $\delta < \rho$ and $y = x - x_o$ and $F = {}_{x_o}f$; then for $x \in U$ and $h \ne 0$ in k with $(x + h \cdot e_i) \in U$,

$$(_{x_o}f(x + h \cdot e_i) - {}_{x_o}f(x))/h$$
$$= \sum_{\alpha\in\mathbb{N}^n} f_\alpha \cdot ((y + h \cdot e_i)^\alpha - y^\alpha)/h$$
$$= \sum_{\alpha\in\mathbb{N}^n} f_\alpha \cdot y^{\alpha'} \cdot (\sum_{j=1}^{\alpha_i} \binom{\alpha_i}{j} \cdot y^{(\alpha_i-j)} \cdot h^{(j-1)})$$

where $\alpha' \in \mathbb{N}^n$ is defined as follows: $\alpha'_j = \alpha_j$ for $j \ne i$ and $\alpha'_i = 0$. Now

$$|y^{\alpha'} \cdot \sum_{j=1}^{\alpha_i} \binom{\alpha_i}{j} \cdot y_i^{(\alpha_i-j)} \cdot h^{(j-1)}|_k$$
$$\le |y|_k^{\alpha'} \cdot \sum_{j=1}^{\alpha_i} \binom{\alpha_i}{j} \cdot |y_i|_k^{(\alpha_i-j)} \cdot |h|_k^{(j-1)}$$
$$\le |y|_k^{\alpha'} \cdot (((|y_i|_k + |h|_k)^{\alpha_i} - |y_i|_k^{\alpha_i})/|h|_k)$$
$$= |y|_k^{\alpha'} \cdot \alpha_i \cdot (|y_i|_k + t(\alpha_i))^{(\alpha_i-1)}$$

for some $t(\alpha_i)$ with $0 \le t(\alpha_i) \le |h|_k$ (by the mean value theorem of the Differential Calculus over $\mathbb{R}$). Thus if h is such that $|y|_k = \rho_1$ and $(|y|_k + |h|_k) = \rho_2$ are both less than or equal to $\rho' < \rho$, the series $\sum_{\alpha\in\mathbb{N}^n} \alpha_i \cdot |f_\alpha|_k(|y|_k +$

$t(\alpha))^{(|\alpha|-1)}$ converges. Thus $D_i f$ is convergent with $\rho(D_i f) \geq \rho$. Hence given an $\epsilon > 0$ there is an $N \in \mathbb{N}$ such that $|\sum_{\alpha \in \mathbb{N}^n, |\alpha| \geq N} f_\alpha \cdot (((y+h \cdot e_i)^\alpha - y^\alpha)/h)| \leq \epsilon$ for all h sufficiently small. Now one has

$$|\{((_{x_o} f(x + h \cdot e_i) -_{x_o} f(x))/h) - \sum_{\alpha \in \mathbb{N}^n}(\alpha_i + 1) \cdot f_\alpha \cdot y^\alpha| - k\}$$
$$= |((F_N(y + h \cdot e_i) - F_N(y))/h) - \sum_{\alpha \in \mathbb{N}^n, |a| \leq N}(\alpha_i + 1) \cdot f_\alpha \cdot y^\alpha|_k + \epsilon$$

where

$$F_N(y) = \sum_{\alpha \in \mathbb{N}^n, |\alpha| \leq (N+1)} f_\alpha \cdot (x - x_o)^\alpha.$$

F_N is a polynomial in y for which it is easily seen that

$$|((F_N(y + h \cdot e_i) - F_N(y))/h) - \sum_{\alpha \in \mathbb{N}^n, |a| \leq N}(\alpha_i + 1) \cdot f_{\alpha + <i>} \cdot y^\alpha|$$

can be made as small as we want by making $|h|_k$ suitably small. This combined with the fact that the series $\sum_{\alpha \in \mathbb{N}^n}(\alpha_i + 1)f_{\alpha + <i>} y^\alpha$ converges to $_{x_o}(D_i f)(x)$ yields the proposition.

2.37. Definitions. Let U be an open set in k^n and $f : U \to k$ be an analytic function. Then for $1 \leq i \leq n$, the limit $\lim_{h \to 0}\{(f(x + h \cdot e_i) - f(x))/h\}$ exists for every $x \in U$. We call the limit the *partial derivative of f with respect to x_i at x* and denote it $(\partial f/\partial x_i)(x)$. The function $\partial f/\partial x_i$ (also denoted $D_i f$) which assigns to each $x \in U$, $(\partial f/\partial x_i)(x)$ is an analytic function on U: this is what Proposition 2.34 asserts. If $\mathcal{A}(U)$ denotes the algebra of analytic functions on U the k-linear endomorphism of $\mathcal{A}(U)$ which assigns to f in $\mathcal{A}(U)$, the function $\partial f/\partial x_i$ is denoted D_i or $\partial/\partial x_i$. One has for f, $g \in \mathcal{A}(U)$,

$$D_i(f \cdot g) = D_i(f) \cdot g + f \cdot D_i(g).$$

We also write $D_i f$ for $D_i(f)$. The endomorphisms $\{D_i | 1 \leq i \leq n\}$ commute with one another (exercise). Define for $\alpha \in \mathbb{N}^n$, D^α as the endomorphism $D_1^{\alpha_1} \circ D_2^{\alpha_2} \cdots \circ D_n^{\alpha_n}$. In the sequel we will denote $D^\alpha \circ D^\beta$ also by $D^\alpha \cdot D^\beta$. We have then $D^\alpha \cdot D^\beta = D^\beta \cdot D^\alpha = D^{(\alpha + \beta)}$. Note that $D^{<i>} = D_i$. For $\alpha \in \mathbb{N}^n$, $D^\alpha(f)$ (also denoted $D^\alpha f$ or $\partial^\alpha f/\partial x^\alpha$) is the αth *partial derivative of f* and is a *partial derivative of order* $|\alpha|$. The theorem below by induction follows from 2.34 (by induction on $|\alpha|$).

2.38. Theorem. *Let U be an open set in k^n and f a k-valued analytic function on U. Then f admits partial derivatives of all orders in U and they are analytic in U. Let $x_o \in U$ and let $T_{x_o} f$ be the formal power series $\sum_{\alpha \in \mathbb{N}^n}(D^\alpha f(x_o)/\alpha!) \cdot X^\alpha$. Then $T_{x_o} f$ is convergent and $f = _{x_o} T_{x_o} f$ in an open set V with $x_o \in V \subset U$.*

$T_{x_o} f$ is the *Taylor Series of f at x_o*.

2.39. Remarks. (**1**) If f is a convergent power series, then we have for the analytic function $F = _{x_o} f$ on $B_{\rho(f)}(x_o)$, $f = T_{x_o} F$. Thus f is uniquely determined by the function $_{x_o} f$ in a neighbourhood of x_o.

(**2**) In case $k = \mathbb{C}$, the existence of the first partial derivatives of a function implies its analyticity. One deduces easily from the Cauchy integral formula the following result:

Let f be a complex analytic function on an open set Ω in $\mathbb{C}^n$. Then for any compact $K \subset \Omega$, there are constants $C(K), d(K) > 0$ such that

$$|D^\alpha f(x)|/\alpha! \leq C(K) \cdot d(K)^{|\alpha|}$$

for all $x \in K$. The same holds for analytic functions over $\mathbb{R}$ (cf. 2.30 (7)) Over $\mathbb{R}$ (and $\mathbb{C}$) one has a converse to this:

A function f on an open set Ω in $\mathbb{R}^n$ is analytic f is C^∞ and for every compact $K \subset \Omega$ there are constants $C(K), d(k) > 0$ such that for all $x \in K$, $|D^\alpha f(x)|/\alpha! \leq C(K) \cdot d(K)^{|\alpha|}$.

Such an assertion is false for non-archimedean fields.

The following result is known as the principle of analytic continuation.

2.40. Proposition. *Let Ω be a connected open set of k^n and f an analytic function on Ω. If $f \equiv 0$ on a <u>non-empty</u> open subset W of Ω, $f \equiv 0$ on Ω. (For non-archimedean k the statement is vacuous.)*

2.41. Proof. Let $\Omega' = \{p \in \Omega | \; D^\alpha f(p) = 0 \; \forall \alpha \in \mathbb{N}^n\}$. Evidently Ω' is a closed set. On the other hand if $p \in \Omega'$, as $f = \; _pT_p(f)$ in a neighbourhood V of p, $f \equiv 0$ on V. It follows that $D^\alpha f(q) = 0$ for all $q \in V$ so that $V \subset \Omega'$. Thus Ω' contains an open neighbourhood of every point in it. It follows that Ω' is open and closed in Ω. As $W \subset \Omega'$, Ω' is non-empty, hence equals Ω proving the proposition.

2.42. Proposition. *Let Ω be an open set in k^n. Let $f : \Omega \to k$ be analytic. Define $\varphi : \Omega \to \mathbb{R}^+$ as follows: for $x \in \Omega$,*

$$\varphi(x) = \min(1, \max \{\rho \in \mathbb{R}^+ | \; _x(T_x f) \text{ converges to } f \text{ in } B_\rho(x) \subset \Omega\}).$$

Then φ is continuous.

2.43. Proof. Fix $x \in \Omega$. Then $T_x f$ converges in $B_{\varphi(x)}(x)$. Let $y \in B_{\varphi(x)}(x)$. By 2.27, we see that $\varphi(y) \geq \varphi(x) - d(x, y)$ or $\varphi(x) - \varphi(y) \leq d(x, y)$. Reversing the roles of x and y, we conclude that $\varphi(y) - \varphi(x) \leq d(y, x) = d(x, y)$. Thus $|\varphi(x) - \varphi(y)| \leq d(x, y)$. Hence the proposition.

2.44. We end this chapter with two basic results about analytic functions on k^n: the Inverse Function Theorem and the existence theorem for the initial value problem for ordinary differential equations. Both theorems are local in character.

$\Omega \subset k^n$ be open set and $F = (f_1, f_2, \cdots, f_m) : \Omega \to k^m$ an analytic map. The *Jacobian matrix* $J(F)(p)$ of F at $p \in \Omega$ is the $(m \times n)$ matrix whose $(i, j)^{\text{th}}$ $(1 \leq i \leq m, 1 \leq j \leq n)$ entry is $(\partial f_i/\partial x_j)(p)$. We also write dF_p for

$J(F)(p)$ and use both also to denote the linear map of k^n in k^m defined by $J(F)(p)$. With this notation, we have:

2.45. Inverse Function Theorem. *Let Ω be an open set in k^n and $F = (f_1, f_2, \cdots, f_n) : \Omega \to k^n$ be an analytic map. Let $x_o \in \Omega$ be a point such that the Jacobian matrix $J(F)_{x_o}$ is invertible. Then F is locally invertible at x_o i.e., there is an open set $V \subset \Omega$ with $x_o \in V$ such that the following holds: the restriction $F|_V$ of F to V is injective, $F(V) = U$ is open in k^n and $(F|_V)^{-1} : U \to V$ is analytic.*

2.46. Proof. Suppose that W is an open set in k^n and $G : W \to k^n$ is an analytic map which is locally invertible at a point $y_o \in W$. We note that if $G(y_o) = x_o$ (resp. $y_o = F(x_o)$), then $F \circ G$ (defined on $G^{-1}(U)$) (resp. $G \circ F$ (defined on $F^{-1}(W)$)) is locally invertible at y_o (resp. x_o) if and only if F is locally invertible at x_o (cf. 2.32). For $v \in k^n$, let $L_v : k^n \to k^n$ denote the translation by v: for $x \in k^n$, $L_v(x) = x + v$. Then L_v is analytic with L_{-v} as its inverse. It is thus locally invertible at every point of k^n. We conclude thus that F is locally invertible if and only if $F' = L_{-F(x_o)} \circ F \circ L_{x_o}$ is locally invertible at 0. The Jacobian matrix of F' at 0 is the same as the Jacobian matrix of F at x_o. Replacing F by F' we assume (as we may) that $x_o = 0$ and $F(0) = 0$.

As linear maps are analytic, composing F with the inverse of the linear *automorphism* of k^n given by the Jacobian matrix, we see that we may assume that the Jacobian matrix of F at 0 is the identity matrix.

Now for $1 \leq r \leq n$ define $\Phi^r = \{\Phi_i^r \mid 1 \leq i \leq n\} : U \to k^n$ as follows: for $1 \leq i \leq r$, $\Phi_i^r = f_i$ and for $j > r$, $\Phi_j^r(x) = x_j$. Assuming now that Φ_1 is locally invertible, we will prove by induction on r that if Φ^i is locally invertible for an integer i with $1 \leq i < n$, then $\Phi^{(i+1)}$ is locally invertible. In fact let V_i be an open neighbourhood of 0 contained in U such that $\Phi^i(V_i) = U_i$ is an open set in U and $\Phi^i|_{V_i}$ admits an analytic inverse G^i. Consider now the analytic function $H = \{h_j \mid 1 \leq j \leq n\} : V_i \to k^n$ defined as follows: $h_j(x) = x_j$ for $j \neq (i+1)$ and $h_{(i+1)}(x) = f_{(i+1)}(G^i(x))$. Then one sees that $\Phi^{(i+1)} = H \circ G^i$. Now by renumbering the coordinates we may assume that $h_j(x) = x_j$ for $j \neq 1$ and the local invertibility of h will follow from the start of the induction. Thus $\Phi^{(i+1)}$ is locally invertible. Since $F = \Phi^n$, the theorem is proved once we can start the induction which is guaranteed (in view of Theorem 2.30) by

2.47. Proposition. *Let $f \in k\{\{X\}\}$ be such that $f_0 = 0$ and $f_{<i>} = \delta_{1i}$ for $1 \leq i \leq n$. Then there is a $g \in k\{\{X\}\}$ such that $g_0 = 0$, $f(g, X_2, X_3, \cdots, X_n) = X_1$ and $g(f, X_2, X_3, \cdots, X_n) = X_1$.*

2.48. Proof. For $\alpha \in \mathbb{N}^n$ with $|\alpha| \geq 2$ set $u_\alpha = -f_\alpha$ so that we have $f = X_1 - \sum_{\alpha \in \mathbb{N}^n, |\alpha| \geq 2} u_\alpha \cdot X^\alpha$. We will first look for a formal power series g

with

$$f(g, X_2, X_3, \cdots, X_n) = X_1 \qquad (*)$$

The left hand side of $(*)$ is

$$\sum_{\alpha \in \mathbb{N}^n, \alpha \neq 0} g_\alpha X^\alpha - \sum_{\alpha \in \mathbb{N}^n, |\alpha| \geq 2} u_\alpha \cdot g^{\alpha_1} \cdot X^{\alpha'}$$

where for $\alpha \in \mathbb{N}^n$, α' is the element in $\mathbb{N}^n$ with $\alpha'_1 = 0$ and $\alpha'_i = \alpha_i$ for $i \geq 2$. Equating coefficients of the $X_i = X^{<i>}$ in $(*)$ we find that $g_{<1>} = 1$ and $g_{<i>} = 0$ for $i > 1$. For $\alpha \in \mathbb{N}^n$ with $|\alpha| \geq 2$, again equating coefficients of X^α in the two sides of $(*)$, we get

$$g_\alpha - \sum_{\beta \in \mathbb{N}^n, |\beta| \geq 2, \beta' \leq \alpha'} u_\beta \cdot (g^{\beta_1})_{\alpha - \beta'} = 0$$

or

$$g_\alpha = \sum_{\beta \in \mathbb{N}^n, |\beta| \geq 2, \beta' \leq \alpha'} u_\beta \cdot (g^{\beta_1})_{\alpha - \beta'}$$

which enables us to define the g_α by induction on $|\alpha|$ as is explained below. The coefficient of X^γ in g^{β_1} is 0 for $|\gamma| \leq \beta_1$ as $g_0 = 0$. Now the coefficient of $X^{(\alpha - \beta')}$ in g^{β_1} is a sum of monomials of the form $\prod_{j=0}^{\beta_1} g_{\gamma(j)}$ with $\sum_{j=0}^{\beta_1} \gamma(j) = (\alpha - \beta')$. It follows that $\gamma(j) \leq \alpha$ and if equality occurs, we have $\beta_1 = 1 = \alpha_1$ and $\beta' = 0$, a contradiction since $|\beta| \geq 2$. Thus we see that $\gamma(j) < \alpha$ for all $1 \leq j \leq \beta_1$. Next we note that the coefficient of X^γ in (g^{β_1}) is zero if $|\gamma| \leq \beta_1$ and thus if it is to be non-zero, we must have $\beta_1 \leq |\alpha - \beta'|$ leading to $|\beta| \leq |\alpha|$. Now if $|\beta| = |\alpha|$, one has $\beta_1 = |(\alpha - \beta')|$. On the other hand if $\{\gamma(j)| 1 \leq j \leq \beta_1\}$ is such that $\sum_{j=1}^{\beta_1} \gamma(j) = (\alpha - \beta')$, one must necessarily have $|\gamma(j)| = 1$ for every $1 \leq j \leq \beta_1$. But if $\prod_{j=1}^{\beta_1} g_{\gamma(j)} \neq 0$, this implies that $\gamma(j) = < 1 >$ for every j and hence $\beta_1 = \alpha_1$ and we conclude that $\beta = \alpha$ if $|\beta| = |\alpha|$. Also when $\beta = \alpha$, $u_\beta \cdot (g^{\beta_1})_{(a-\beta')} = u_\alpha \cdot (g^{\alpha_1})_{\alpha_1 \cdot <1>} = u_\alpha$. We conclude thus that

$$g_\alpha = u_\alpha + \sum_{\beta \in \mathbb{N}^n, |\beta| < |\alpha|} u_\beta \cdot Q_\beta \qquad (**)$$

where Q_β is a sum of monomials of the form $g_{\gamma(1)} \cdot g_{\gamma(2)} \cdots g_{\gamma(\beta_1)}$ with $\gamma(j) < \alpha$ and $\sum_{j=0}^{\beta_1} \gamma(j) = (\alpha - \beta')$. When $|\alpha| = 2$, there is no $\beta \in \mathbb{N}^n$ with $2 \leq |\beta| < |\alpha|$ so that from $(**)$ we find that $\gamma_\alpha = u_\alpha$. We will now prove by induction on $|\alpha|$ the following. For every $\alpha \in \mathbb{N}^n$ with $|\alpha| \geq 2$ there is a polynomial P_α in $\mathbb{Z}[\{T_\beta | \beta \in \mathbb{N}^n, 2 \leq |\beta| \leq |\alpha|\}]$ with *non-negative integral* coefficients such that

$$g_\alpha = u_\alpha + P_\alpha(\{u_\beta | \beta \in \mathbb{N}^n, 2 \leq |\beta| \leq |\alpha|\}) \qquad (***)$$

When $|\alpha| = 2$, $P_\alpha = 0$, so we can start the induction. Once the result is established for $q = |\alpha| \geq 2$, it is clear from $(**)$ that it holds for $|\alpha| = (q+1)$. Thus we have shown that there is a $g \in k[[X]]$, such that $f(g, X_2, X_3, \cdots, X_n) = X$. Now g has the same form as f and hence there is a $f' \in k[[X]]$ such that $g(f', X_2, X_3, \cdots, X_n) = X$. Now $f = f(g(f', X_2, X_3, \cdots, X_n), X_2, X_3, \cdots,$

$X_n) = f'$ so that we have $f = f'$. In order to establish Proposition 2.44, it suffices to show that $g \in k\{\{X\}\}$. We may assume after a suitable scaling (note that $f_0 = 0$) that $|u_\alpha| \leq 1$ for all $\alpha \in \mathbb{N}^n$.

Let $F \in \mathbb{R}\{\{X\}\}$ be the series $X_1 - \sum_{\alpha \in \mathbb{N}^n, |\alpha| \geq 2} X^\alpha = 1 + 2 \cdot X_1 + (\sum_{i=2}^n X_i) - (\prod_{i=1}^n (1 - X_i)^{-1})$ so that setting $(\sum_{i=2}^n X_i) = A$ and $\prod_{i=2}^n (1 - X_i)^{-1} = B$, we get

$$2 \cdot X_1^2 - (F - A - 1) \cdot X_1 + (F + A + B - 1) = 0;$$

it follows that

$$X_1 = ((F - A - 1) + \{(F - A - 1)^2 - 8 \cdot (F + A + B - 1)\}^{1/2})/4.$$

All the operations on the right hand side take place in the ring of convergent power series over $\mathbb{R}$. Note that $(F - A - 1)^2 - 8 \cdot (F + A + B - 1)$ is of the form $1 + C$ with $C \in \mathfrak{m}_o$ so that it can be substituted in the binomial series of exponent $1/2$. Now let G be the convergent power series

$$((X_1 - A - 1) + \{(X_1 - A - 1)^2 - 8 \cdot (X_1 + A + B - 1)\}^{1/2})/4;$$

then $X_1 = G(F, X_2, X_3, \cdots, X_n)$. If we set $U_\alpha = -F_\alpha$ for $|\alpha| \geq 2$, then for $\alpha \in \mathbb{N}$ with $|\alpha| \geq 2$ in view of $(* * *)$ (applied to $F \in \mathbb{R}\{\{X\}\}$ instead of f),

$$G_\alpha = (U_\alpha + P_\alpha(\{U_\beta | \beta \in \mathbb{N}^n, 2 \leq |\beta| < |\alpha|\})).$$

Now since the P_α are polynomials with *non-negative integral coefficients*, $|g_\alpha|_k \leq G_\alpha$ for all $\alpha \in \mathbb{N}^n$ with $|\alpha| \geq 2$. Since G is convergent so is g. Hence the proposition.

2.49. Implicit Function Theorem. *Let Ω be an open set in $k^m \times k^n$ and $F = \{f_i | 1 \leq i \leq m\} : \Omega \to k^m$ be an analytic function. Let $(y_o, x_o) = (\{y_{oi}\}_{1 \leq i \leq m}, \{x_{oi}\}_{1 \leq i \leq n})$ be a point in Ω such that the sub-matrix $\{(\partial f_i / \partial y_j)((y_o, x_o))\}_{1 \leq i,j \leq m}$ at (y_o, x_o) of $J(F)(y_o, x_o)$ is non-singular. Then there are open neighbourhood U of x_o in k^n, V of y_o in k^m with $V \times U \subset \Omega$ and an analytic function $\Phi : U \to V$ such that for $(y, x) \in V \times U$, $F(y, x) = F(y_o, x_o)$ if and only if $y = \Phi(x)$.*

2.50. Proof. Consider the analytic map $\tilde{F} : \Omega \to k^{(m+n)}$ defined by setting for $(y, x) \in \Omega$, $\tilde{F}(y, x) = (F(y, x), x)$. Then the Jacobian matrix of $\tilde{F}$ at (y_o, x_o) is easily seen to be non-singular. It follows from the Inverse Function Theorem that there is an open neighbourhood of (y_o, x_o) (which may be taken to be of the form $V \times U$) such that $\tilde{F}$ maps $V \times U$ homeomorphically on an open neighbourhood $\tilde{\Omega}'$ of $(F(y_o), x_o)$ in $k^{(m+n)}$. Define now $\Phi : U \to V$ by setting for $x \in U$, $(\Phi(x), x) = (\tilde{F})^{-1}(F(y_o, x_o), x)$. It is then clear that U, V, Φ satisfy the requirements of the theorem.

2.51. Remarks. (1) The theorem says that near the point (y_o, x_o), the set $F^{-1}(F(y_o, x_o))$ looks like the graph of an analytic function from an open set in k^n into k^m and is hence homeomorphic to an open set in k^n.

(**2**) One can reformulate the theorem as follows. Let Ω be an open set in $k^{(m+n)}$ and $F : \Omega \to k^m$ be an analytic function. Suppose that $x_o = \{x_{oi}|\ 1 \leq i \leq (m+n)\}$ is a point such that $J(F)$ has rank m. Then there is a subset $I \subset [1, m+n]$ of cardinality n, a neighbourhood V of $\{x_{oi}\}_{i \in I}$ in $\mathbb{R}^n$, a neighbourhood U of $\{x_{oj}\}_{j \notin I}$ in $\mathbb{R}^m$ and an analytic function $\Phi = \{\Phi_i\}_{i \notin I} : V \to U$ such that the following holds. Set $J = [1, (m+n)] - I$. Let $W = \{x \in \Omega|\ \{x_i\}_{i \in I} \in V, \{x_j\}_{j \in J}$. Then for $x \in W$, $F(x) = F(x_o)$ if and only if $x_j = \Phi_j(\{x_i\}_{i \in I})$ for all $j \in J$: Choose $J = \{r_1, r_2, \cdots, r_m\}$ such that the sub-matrix $\partial f_i / \partial x_{r_j}$ is non-singular at $(y - o, x_o)$. Renumbering the coordinates so that x_{r_i} become $\{x_i$ for $1 \leq i \leq m$, we recover the formulation as in Theorem 2.48.

(**3**) The reformulation tells us the following: if some $m \times m$ sub-matrix $\{(\partial f_i / \partial x_{r_j})\ (x_o)\}_{1 \leq i,j \leq m}$ of the Jacobian is non-singular, then every coordinate of points on $\{x \in \Omega| F(x) = F(x_o)\}$ near $(yo, x_o$ is an analytic function of the coordinates $\{x_i|\ i \in I\}$.

2.52. Rank Theorem. *Let Ω be an open set in k^n and $F = \{f_i\}_{1 \leq i \leq m} : \Omega \to k^m$ be analytic function. Suppose that rank $dF_x = r$ is an integer independent of $x \in \Omega$. Then for any $a \in \Omega$ there exist*
(i) an open neighbourhood U of a,
(ii) an open neighbourhood V of $b = F(a)$,
(iii) cubes Q_1, Q_2 (cf. 1.92) around $a \in k^n$ and $b \in k^m$ respectively,
(iv) analytic diffeomorphisms $u_1 : Q_1 \to U, u_2 : V \to Q_2$
such that we have, setting $\Phi = u_2^{-1} \circ F \circ u_1$,

$$\Phi(x_1, x_2, \cdots, x_n) = (x_1, x_2, \cdots, x_r, 0, \cdots, 0).$$

2.53. Proof. Note that $r \leq \min(m, n)$. By composing with suitable translations and linear maps of k^n and k^m we can assume that $a = 0, b = 0$ and $dF_a(\xi_1, \xi_2, \cdots, \xi_n) = (\xi_1, \xi_2, \cdots, \xi_r, 0, \cdots, 0)$. Consider the map $u : \Omega \to k^n$ defined by setting for $x \in \Omega$,

$$u(x) = (f_1(x), f_2(x), \cdots, f_r(x), x_{r+1}, \cdots, x_n).$$

Then $du_0 = $ identity, hence by the Inverse Function Theorem there exists a neighbourhood U of 0 and a cube $Q_1 \subset \Omega$ such that $u|_U : U \to Q_1$ is an analytic diffeomorphism. Let $u^{-1}|Q_1 = u_1$. Clearly for $y \in Q_1$, $f(u_1(y)) = (y_1, y_2, \cdots, y_r, \Phi_{r+1}(y), \cdots, \Phi_m(y))$ where the $\{\Phi_i|\ r+1 \leq i \leq m\}$ are analytic functions on Ω. If $\psi(y) = F(u_1(y))$, obviously rank $d\psi_y = r$ for all $y \in Q_1$ and hence

$$(\partial \Phi_j / \partial y_k)(y) = 0, \ \forall y \in Q_1 \text{ and } j, k > r.$$

It follows that for $y \in Q_1$

$$\Phi_j(y) = \Phi_j(y_1, y_2, \cdots, y_r) \ for \ j > r,$$

i.e., the $\Phi_j, j > r$ are independent of the last $n - r$ variables. Now $Q_1 = P \times P'$ where P and P' are cubes in k^r and k^{n-r} respectively. Define $u_2 : P \times k^{m-r} \to P \times k^{m-r}$ as follows: for $\underline{y}' = (y_1, y_2, \cdots, y_r)) \in P$ and $\underline{z}(= (y_{r+1}, y_{r+2}, \cdots, y_m)) \in k^{m-r}$,

$$u_2(\underline{y}') = (\underline{y}', y_{r+1} - \Phi_{r+1}(\underline{y}'), y_{r+2} - \Phi_{r+2}(\underline{y}'), \cdots, y_m - \Phi_m(\underline{y}')).$$

Then u_2 is an analytic diffeomorphism of $P \times k^{m-r}$ on itself with

$$u_2^{-1}((\underline{y}'), \underline{z}) = (\underline{y}', y_{r+1} - \Phi_{r+1}(\underline{y}'), y_{r+2} - \Phi_{r+2}(\underline{y}'), \cdots, y_m + \Phi_m(\underline{y}')).$$

as its inverse. Let $Q_2 \subset k^m$ be a cube with $u_2(\psi(Q_1)) \subset Q_2$; then $\Phi = u_2^{-1} \circ F \circ u_1(x_1, x_2, \cdots, x_n) = (x_1, x_2, \cdots, x_r, 0, \cdots, 0)$.

We now state and prove the theorem on the existence (and uniqueness) of a solution to the initial value problem for a family of ordinary analytic differential equations (depending analytically on a parameter).

2.54. Theorem. *Let Ω (resp. I, resp. U) be an open set in k^m (resp. k, resp. k^n). Let $F : \Omega \times I \times U \to k^n$ and $\mathcal{I} : \Omega \to U$ be analytic functions. Then given a point $t_o \in I$ and a relatively compact subset K of Ω, there is an open neighbourhood Ω_o of K in Ω, a $\delta > 0$ with $I_\delta(t_o) \subset I$ and an analytic function $G : \Omega_o \times I_\delta(t_o) \to U$ such that*
(i) $(\partial G/\partial t)(\xi, t) = F(\xi, t, G(\xi, t))$ and
(ii) $G(\xi, t_o) = \mathcal{I}(\xi), \ \forall \ \xi \in \Omega_o$.
Moreover, if G' is a U-valued analytic function on $\Omega'_o \times I_{\delta'}(t_o)$ where Ω'_o is an open neighbourhood of K in Ω and δ' is > 0, satisfying (i) and (ii), then there is a $\delta_1 \leq \inf(\delta, \delta')$ and an open neighbourhood of K contained in $\Omega_o \cap \Omega'_o$ such that $G = G'$ on $\Omega_1 \times I_{\delta_1}(t_o)$.

2.55. Proof of 2.54: Step 1. Set $\tilde{U} = U \times k$. Define $\tilde{\mathcal{I}} : \Omega \to \tilde{U}$ by $\tilde{\mathcal{I}}(\omega) = (\mathcal{I}(\omega), t_o)$ and $\tilde{F} : \Omega \times I \times (U \times k) \to k^{(m+1)}$ by $\tilde{\mathcal{I}}(\omega) = (\mathcal{I}(\omega), t_o)$ and $\tilde{F}(\omega, t, \{y_i | 1 \leq i \leq (m+1)\}) = (F(\omega, y_{m+1}, \{y_i | 1 \leq i \leq m\}), 1)$. Then one sees that $\tilde{G} = (G, t)$ is a solution to $(\partial \tilde{G}/\partial t)(\omega, t) = \tilde{F}(\omega, \tilde{G})$ with $\tilde{\mathcal{I}}(\omega) = \tilde{G}(\omega, t_o)$ iff G is a solution to $(\partial G/\partial t)(\omega, t) = F(\omega, t, G(\xi, t))$ with $G(\omega, t_o) = \mathcal{I}(\omega)$. Thus we may assume that F is independent of t.

2.56. Proof of 2.54: Step 2. If the theorem holds when $U = k^m$ it holds for any open set U as well. This is because starting with a $G : \Omega_o \times I_\delta(t_o) \to k^m$ satisfying (i) and (ii), by replacing Ω_o by a relatively compact open subset and δ by a smaller constant we can ensure (in view of (ii) and the uniform continuity of G on compact sets), that $G(\Omega_o \times I_\delta(t_o)) \subset U$. Next, we may assume that $\mathcal{I} \equiv 0$: if G_1 satisfies (i) and (ii) with $\mathcal{I} \equiv 0$, $G = G_1 + \mathcal{I}$ satisfies (i) and (ii). Suppose that we have proved that for any $p \in K$, there is an open disc Ω_p in k^m containing p, a $\delta_p > 0$ and an analytic function

$G_p : \Omega_p \times I_{\delta_p} \to U$ such that
(i) $(\partial G_p / \partial t)(\omega, t) = F(\xi, G_p(\omega, t))$ and
(ii) $G_p(\omega, t_o) = \mathcal{I}(\omega)$ for all $(\omega, t) \in \Omega_p \times I_{\delta_p}(t_o)$.
Since K is compact we can find a finite set S such that $\cup_{p \in S} \Omega_p \supset K$.

2.57. Proof of 2.54: Step 3; Case of Non-Archimedean k.

When k is non-archimedean, we assume as we may, that if $p, q \in S, p \neq q, \Omega_p \cap \Omega_q = \phi$. Assume (replacing δ_p by a smaller number, if necessary) that the Taylor series of G_p at $(p, 0)$ converges uniformly and absolutely in $\Omega_p \times I_{\delta_p}(t_o)$ to G_p. Set $\delta = \inf\{\delta_p | p \in S\}$. We define $G : \Omega_o \times I_\delta(t_o) \to U$ by setting $G|_{\Omega_p \times I_\delta(t_o)} = G_p$ for $p \in S$. G is well defined as the $\{\Omega_p | p \in S\}$ are mutually disjoint. Thus for non-archimedean k, the theorem follows from the special case when K is a point.

2.58. Proof of 2.54: Step 4; The Archimedean Case.

Observe that in the case when k is archimedean, the $\Omega_p, p \in S$ being discs, $\Omega_p \cap \Omega_q$ is connected for any pair of elements $p, q \in S$. Both G_p and G_q are solutions of the differential equation in $(\Omega_p \cap \Omega_q) \times I_\delta$ and in view of the uniqueness assertion in the theorem (for the case of a point) the germ of $G_p - G_q$ at any point of $(\Omega_p \cap \Omega_q) \times \{t_o\}$ is zero. It follows that $G_p - G_q$ is zero on an open subset of $(\Omega_p \cap \Omega_q) \times I_\delta$; and by the principle of analytic continuation is zero on all of $(\Omega_p \cap \Omega_q) \times I_\delta$. Thus $G_p = G_q$ on $(\Omega_p \cap \Omega_q) \times I_\delta$ for every $p, q \in S$. Thus we can define G on $\cup_{p \in S}(\Omega_p \times I_\delta)$ by setting $G(\omega, t) = G_p(\omega, t)$ for $(\omega, t) \in \cup_{p \in S} \Omega_p \times I_\delta$ for any $p \in S$ with $\omega \in \Omega_p$. We are reduced to proving the theorem when K is a point and $\mathcal{I} \equiv 0$, which is the final step.

The special case when K is a point and $\mathcal{I} \equiv 0$ is easily seen to be equivalent to the following result about convergent power series.

2.59. Theorem.

Let $f = \{f_i | 1 \leq i \leq n\}$ be an n-tuple of power series in $m + n$ variables $\{\{Z_i | 1 \leq i \leq m\}, \{Y_j | 1 \leq j \leq n\}\}$. Then there is a unique n-tuple of formal power series $g_f = \{g_{fi} | 1 \leq i \leq n\}$ in the variables $\{\{Z_i | 1 \leq i \leq m\}, T\}$ such that
(i) *$g_{fi(\alpha,0)} = 0$ for all $\alpha \in \mathbb{N}^n$ and $1 \leq i \leq n$ a and*
(ii)*$D_T(g_{fi}) = f_i(Z, g_f(Z, T))$.*
Moreover if the f_i, $1 \leq i \leq n$ are convergent so are the g_{fi}.

2.60. Proof of 2.58.

We are looking for an n-tuple of power series $g_f = \{g_{fi}(Z, T) | 1 \leq i \leq n\}$ in the variables Z, T with $g_{fi(\alpha,0)} = 0$ for all $\alpha \in \mathbb{N}^m$ such that

$$D_T(g_{fi}) = f_i(Z, g_f(Z, T)). \qquad (*)$$

For an $\alpha \in \mathbb{N}^m$ and $r \in \mathbb{N}$, equating coefficients of $Z^\alpha \cdot T^r$ in the two sides of the above equation $(*)$, we have

$$(r+1) \cdot g_{fi}(\alpha, (r+1)) = \sum_{(\beta,\gamma) \in \mathbb{N}^m \times \mathbb{N}^n} f_{i(\beta,\gamma)} \cdot (\prod_{l=1}^{n} g_{fl}^{\gamma_l})(\alpha,r). \tag{1}$$

We introduce the following notation. Let $\mathcal{B}$ be the set of finite subsets of $\mathbb{N}^m \times \mathbb{N}$. For $\gamma \in \mathbb{N}^n$, and $1 \leq l \leq n$ let $\mathcal{C}_{(\gamma,l)}$ be the set of maps of $[1, \gamma_l]$ in $\mathbb{N}^m \times \mathbb{N}$ and finally for $(\beta,\gamma) \in \mathbb{N}^m \times \mathbb{N}^n$ and $(\alpha, r) \in \mathbb{N}^m \times \mathbb{N}$ let $\mathcal{D}((\beta,\gamma),(\alpha,r)) =$

$$\{u = \{u_l | 1 \leq l \leq n\} \in \prod_{l=1}^{n} \mathcal{C}_{\gamma,l} | \ (\beta,0) + \sum_{1 \leq l \leq n} \sum_{q=1}^{\gamma_l} u_l(q) = (\alpha,r)\}$$

With this notation we see that (1) can be rewritten as

$$(r+1)g_{fi(\alpha,r+1)} = \sum_{(\beta,\gamma) \in \mathbb{N}^m \times \mathbb{N}^n} f_{i(\beta,\gamma)} \cdot \prod_{u \in \mathcal{D}((\beta,\gamma),(\alpha,r))} g_{flu_l(q)} \tag{2}$$

Set $g_{fi(\alpha,r)}^* = r! \cdot g_{fi(a,r)}$. Then multiplying both sides of (2) by $r!$, we have, setting $u_l(q) = (\xi_{lq}, r_{lq})$,

$$g_{fi(\alpha,r+1)}^* = \sum_{(\beta,\gamma) \in \mathbb{N}^{(m+n)}} f_{i(\beta,\gamma)} \cdot \prod_{u \in \mathcal{D}((\beta,\gamma),(\alpha,r))} r!/(\prod_{l=1}^{n} \prod_{q=1}^{\gamma_l} r_{lq}!) \cdot g_{flu(q)}^* \tag{3}$$

Note: $r!/(\prod_{l=1}^{n} \cdot \prod_{q=1}^{\gamma_l} r_{lq}!)$ is an integer > 0 as $\sum_{l=1}^{n} \sum_{q=1}^{\gamma_l} r_{lq} = r$.

2.61. Proof of 2.58 (Continued): The Case k is Non-Archimedean.
Let $C > 0$ be a constant such that $|f_{i(\beta,\gamma)}|_k \leq C^{|\beta|+|\gamma|}$ for all $(\beta,\gamma) \in \mathbb{N}^m \times \mathbb{N}^n$. We assume $C \geq p$ where p is the residue field characteristic of k. We will now show that $|g_{fi(\alpha,r)}^*|_k \leq C^{|\alpha|+(r-1)}$ for all $\alpha \in \mathbb{N}^m$ and all $r \in \mathbb{N}$. This is by induction on r. This holds at the start of the induction when $r = 0$ since $g_{fi(\alpha,0)} = 0$ for all $\alpha \in \mathbb{N}^m$. Assume the claim proved for all $r \leq q$ and all $\alpha \in \mathbb{N}^m$ and $r = (q+1)$ and all $\alpha \in \mathbb{N}^m$ with $|\alpha| \leq s$. Now by (3), in view of the fact that $|z|_k \leq 1$ for all $z \in \mathbb{Z}$ and for $x, y \in k, |x+y|_k \leq \max(|x|_k, |y|_k)$, we see that we have for $\alpha \in \mathbb{N}^m$ with $|\alpha| = s+1$, $|g_{fi(\alpha,q+1)}^*|_k$ is less than or equal to

$$\max\{|f_{i(\beta,\gamma)}|_k \cdot (\prod_{l=1}^{n} \prod_{u=1}^{\gamma_l} | \ g_{flu(q)}^*|_k)| \ (\beta,\gamma) \in \mathbb{N}^m \times \mathbb{N}^n, u \in D((\beta,\gamma),(\alpha,r))\}$$

By the induction hypothesis $|g_{flu(q)}^*|_k \leq C^{\sum_{l=1}^{n} \sum_{q=1}^{\gamma_l} \xi_{lq}+(r_{lq}-1)}$ so that $|g_{fi(\alpha,q+1)}|_k^* \leq C^{|\alpha|+r}$ and hence

$$|g_{fi(\alpha,r)}|_k = |1/r!|_k \cdot |g_{fi(\alpha,r)}^*|_k \leq |1/r!|_k \cdot C^{|\alpha|+(r-1)}$$

Now $|1/r!|_k \cdot p^{[r/p]+[r/p^2]\cdots[r/p^t]\cdots} \leq p^{r/(p-1)} \leq p^r$ so that

$$|g_{fi(\alpha,r)}|_k \leq C^{|\alpha|+2r-1}.$$

Hence the power series g_{fi}, $1 \leq i \leq n$, are convergent.

2.62. Proof of 2.58 (Continued): The Case k is Archimedean. We argue a little differently. From (3) of 2.59 we deduce easily by an induction on $|(\alpha, r)|$ the following:

Assertion: *For every $1 \leq i \leq n$ and $(\alpha, r) \in \mathbb{N}^m \times \mathbb{N}^n$, there is a polynomial $P_{i(\alpha,r)}$ in $\mathbb{Z}[\{X_{j(\beta,\gamma)}| 1 \leq j \leq n, |\beta| \leq |\alpha|, |\gamma| \leq r\}]$ with all coefficients non-negative, such that for any $f = \{f_i | 1 \leq i \leq n\}$, for the formal solution $g_f = \{g_{fi}| 1 \leq i \leq n\}$ of $DT(g_f) = f(Z, g_f)$, we have*

$$g^*_{fi(\alpha,r)} = P_{i(\alpha,r)}(\{f_{i(\beta,\gamma)}| 1 \leq j \leq n, |\beta| \leq |\alpha|, |\gamma| \leq r\}).$$

Now, let $C > 0$ be such that $|f_i(\alpha, \beta)|_k \leq C^{|\alpha|+|\beta|}$ for all $(\alpha, \beta) \in \mathbb{N}^m \times \mathbb{N}^n$. Let F_i, $1 \leq i \leq n$ be the (convergent) power series (over $k = \mathbb{R}$ or $\mathbb{C}$) $\sum_{(\alpha,\beta) \in \mathbb{N}^m \times \mathbb{N}^n} C^{|\alpha|+|\beta|+1} Z^\alpha \cdot Y^\beta$ which sums up to the analytic function

$$C^{m+n} \cdot \prod_{i=1}^{m}(1 - C \cdot z_i)^{-1} \cdot \prod_{j=1}^{n}(1 - C \cdot y_j)^{-1}.$$

Consider now the system of differential equations

$$dy_i(z,t)/dt = C^{m+n+1} \prod_{i=1}^{m}(1 - C \cdot z_i)^{-1} \cdot \prod_{j=1}^{n}(1 - C \cdot y_j(t))^{-1} \qquad (*)$$

with the initial condition $y_i(0) = 0$. Now by inspection we see that

$$y_i(t) = C^{-1}[1 - C^{(m+n+2)} \cdot (\prod_{j=1}^{m}(1 - C \cdot z_j)^{-1}) \cdot t]^{1/(m+1)},$$

$1 \leq i \leq n$ which are all analytic functions of $(\{z_i| 1 \leq i \leq n\}, t)$ near $(0,0)$ in $k^m \times k$, is a solution of the system $(*)$. Note that F_i is independent of i. This implies that the formal power series solution g_F of $D_T g_F = F(g_F)$ is convergent. The Assertion yields the following inequality:

$$|g_{fi(\alpha,r)}|_k = |P_{i(\alpha,r)}(f_{j(\beta,\gamma)})|_k \cdot r! \leq P_{i(\alpha,r)}(F_{j(\beta,\gamma)})/r! = |g_{Fi(\alpha,r)}|_k.$$

We have here used the fact that for archimedean k, $|q|_k = q$ for all $q \in \mathbb{N}$. As the g_{Fi} are convergent, so are the g_{fi} proving Theorem 2.55 for archimedean k.

2.63. Remarks. (1) In general, in Theorem 2.54, δ depends on K and $\mathcal{I}(t_o)$. The following example where $k = \mathbb{R}$, $m = 0$ (so that K is a point) $n = 1$ and $I = (-a, a)$ with $a > 1$ illustrates this: the solution to $dy/dt = e^{-y}$ with $\mathcal{I}(z) = u > 0$ is given by $y = \ln(t - t_o + \ln(u))$ so that the solution is valid only in $|t - t_o + \ln(u)| < 1$.

(**2**) When k is archimedean the uniqueness of the germ of a solution combined with the principle of analytic continuation guarantees (in view of the connectedness of $I_\delta(t_o)$) that the solution g in Theorem 2.54 is unique on all of $\Omega_o \times I_\delta(t_o)$.

3. Analytic Manifolds

In this chapter we outline some of the theory of analytic manifolds over a local field k. Anyone familiar with real analytic manifolds will see that this outline is a simple carry over of basic concepts and results on real analytic manifolds to cover the non-archimedean case. We do not touch upon the deeper results on real or complex analytic manifolds. In particular we do not deal with the topology of real or complex manifolds which is a fascinating subject. The topology of analytic manifolds over a non-archimedean field, on the other hand, is far from interesting: by a theorem of Serre, every paracompact analytic manifold over a non-archimedean field is analytically isomorphic to a disjoint union of discs.

Throughout this chapter, unless otherwise specified, k will denote a local field of characteristic 0.

3.1. Recall that a real m-dimensional topological manifold is a Hausdorff topological space M such that every point in M has an open neighbourhood homeomorphic to an open set in $\mathbb{R}^m$. This notion does not give rise to any inconsistencies as an open set in $\mathbb{R}^m$ is not homeomorphic to an open set in $\mathbb{R}^{m'}$ if $m \neq m'$ (this is a deep result due to L E J Brouwer). In contrast, we have for any prime p a homeomorphism h of $\mathbb{Z}_p$ (an open set in $\mathbb{Q}_p$) on $\mathbb{Z}_p^2$ (an open set in $\mathbb{Q}_p^2$) defined as follows: for $x = \sum_{i=0}^{\infty} a_i \cdot p^i$, $a_i \in \mathbb{Z}, 0 \leq a_i < p$ in $\mathbb{Z}_p$, $h(x) = (\sum_{i=0}^{\infty} a_{2i} \cdot p^i, \sum_{i=0}^{\infty} a_{(2i+1)} \cdot p^i)$. However the more delicate notion of a real (or complex) *analytic* manifold does indeed have a counter part over arbitrary local fields. The definition made in the case of $\mathbb{R}$ (or $\mathbb{C}$) is suggested by the implicit function theorem: and that definition carries over to arbitrary k, as the Implicit Function Theorem for analytic functions is available over any local field k (see 2.48) (note that the map h defined above is not analytic). If $F : U_1 \to U_2$ is a bijection of a non-empty open set $U_1 \subset k^{m_1}$ on an open set $U_2 \subset k^{m_2}$ such that both F and F^{-1} are *analytic*, then $m_1 = m_2$.

3.2. Definitions. Let Ω, Ω' be open sets in k^m and $F : \Omega \to \Omega'$ be a bijection. F is an *analytic diffeomorphism* if F as well as its inverse $F^{-1} : \Omega' \to \Omega$ are analytic. In view of the inverse function theorem, a bijection $F = \{f_i\}_{1 \leq m} \Omega \to \Omega'$ is an analytic diffeomorphism if and only if for every $p \in \Omega$, the Jacobian matrix $\{\partial f_i / \partial x_j(p)\}_{1 \leq i,j \leq m}$ is non-singular. For an integer $m \geq 0$, an *m-coordinate chart over k* (or an *m-coordinate system over k* or simply an *m-chart over k*) on a topological space M, is a pair (U, Φ) where U is an open set in M and Φ is a homeomorphism of U onto an open set $U'(= \Phi(U))$ in k^m. Once a chart (U, Φ) is chosen, points in U are often treated as points in $\Phi(U) \subset k^m$ and are thus specified by an m-tuple $\{x_i\}_{1 \leq i \leq m} \in k^m$.

Two m-charts (U, Φ), (V, Ψ) on M are *(analytically) compatible* if the maps $\Phi \circ \Psi^{-1}$ and $\Psi \circ \Phi^{-1}$ are analytic maps of $\Psi(U \cap V)$ in $\Phi(U \cap V)$ and $\Phi(U \cap V)$ in $\Psi(U \cap V)$ respectively; in other words $\Phi \circ \Psi^{-1}$ is an analytic diffeomorphism of $\Psi(U \cap V)$ on $\Phi(U \cap V)$. An *(analytic) m-atlas* on M is a collection $\mathcal{U} = \{(U_i, \Phi_i) | i \in I\}$ of m-charts on M such that $\bigcup_{i \in I} U_i = M$ and, for every i, j in I, (U_i, Φ_i) and (U_j, Φ_j) are analytically compatible. Two atlases $\mathcal{U} = \{(U_i, \Phi_i) | i \in I\}$ and $\mathcal{V} = \{(V_j, \Psi_j) | j \in J\}$ are *equivalent* if $\mathcal{U} \cup \mathcal{V}$ is an atlas (this is indeed an equivalence relation: reflexivity and symmetry are obvious; transitivity is a result of the fact that composites of analytic maps are analytic).

An analytic manifold over k of dimension m (or an analytic m-manifold) is a topological space M together with an equivalence class $\underline{A}$ of m-atlases on it. Clearly an analytic m-manifold $(M, \underline{A})$ is determined by a single atlas on $M \in \underline{A}$. The *dimension* of M is m. In the sequel, we will let M stand for the analytic m-manifold $(M, \underline{A})$. Also we simply say manifold for an analytic m-manifold and in this chapter, if an analytic manifold is denoted by a Roman capital letter, its dimension is denoted by the corresponding lower case letter. Analytic manifolds over $\mathbb{R}$ (resp. $\mathbb{C}$) will also be referred to as real analytic (resp. complex analytic or complex) manifolds.

3.3. We will also assume the underlying topological space of any analytic manifold to be a *countable union of compact subsets*. This is equivalent to assuming that it is second countable. Consequently every open covering of the manifold admits a *countable* sub-cover. Also, this would mean that the underlying topological space is paracompact (cf. 1.69). When k is non-archimedean, this has the implication that any open covering admits a locally finite refinement by *open and closed sets* such that any two distinct sets in the refinement are disjoint. This follows from the fact that every $p \in M$ has a fundamental system of open and closed neighbourhoods. The following result is not difficult to prove.

3.4. Theorem. (Serre [14]) *If k is non-archimedean, an analytic manifold M over k admits an atlas $\{(U_i, \Phi_i) | i \in I\}$ such that $\Phi_i(U_i) = B_1(0)$ and for $i \in I$ and $U_i \cap U_j = \phi$ for $i, j \in I$ with $i \neq j$.*

3.5. Analytic manifolds over $\mathbb{R}$ and $\mathbb{C}$ have no such simple classification. Their study constitutes an elaborate and beautiful theory which is deep and difficult. In this book we touch only upon some simpler aspects of the theory which they share with analytic manifolds over non-archimedean fields. They are adequate for our study of Lie groups.

3.6. Remark. Suppose that k' is a subfield of k containing $\hat{\mathbb{Q}}$. Then an analytic manifold $(M, \underline{A})$ over k carries a natural analytic manifold structure over k'. In fact $\{(U_i, \Phi_i) | i \in I\}$ is in $\underline{A}$, the 'transition' functions $\Phi_i \circ \Phi_j^{-1}$ being analytic functions over k are analytic over k' as well (see 2.30 (8)). This k' analytic manifold will be denoted $R_{k/k'} M$. It is of dimension $m \cdot [k : k']$ where m is the dimension of M over k and $[k : k']$ is the dimension of k as a vector space over k'.

3.7. Admissible Atlases. When k is non-archimedean an *admissible atlas* on an analytic manifold M is an atlas $\{(U_i, \Phi_i) | i \in I\}$ satisfying the following three conditions: (i) $\Phi_i(U_i) = B_1(0)$, (ii) I is countable and (iii) $U_i \cap U_j = \phi$ for $i, j \in I$ with $i \neq j$. Such an atlas exists (Theorem 3.4).
When k is archimedean an atlas $\{(U_i, \Phi_i) | i \in I\}$ is *admissible* if (i) for every $i \in I$, U_i is relatively compact, (ii) I is countable and (iii) the covering $\mathfrak{U} = \{U_i | i \in I\}$ is locally finite. Such a $\mathfrak{U}$ exists since M is paracompact.

3.8. Metric on. An admissible atlas $\mathcal{U} = \{(U_i, \Phi_i) | i \in I\}$ on M enables us to define a metric $d = d_{\mathcal{U}}$ on M. Consider first the case when k is non-archimedean. Then for any $x \in M$, there is a unique $i(x) \in I$ with $x \in U_{i(x)}$. For $x, y \in M$, $d(x, y) = 1$ if $i(x) \neq i(y)$; if $i(x) = i(y) = i$, $d(x, y) = d(\Phi_i(x), \Phi_i(y))$, where on the right hand side of the equation d is the distance in k^m.

In the case when k is archimedean the metric is defined as follows. Let $\mathfrak{V} = \{V_i | i \in I\}$ be a shrinking of the covering $\mathfrak{U} = \{U_i | i \in I\}$ (this means that V_i is open for every $i \in I$, the closure of V_i in M is contained in U_i for every $i \in I$ and $\bigcup_{i \in I} V_i = M$). Then for $x, y \in M$ define $d(x, y) = 1$ if $\{i \in I | x, y \in V_i\} = \phi$ while $d(x, y) = \max\{d(\Phi_i(x), \Phi_i(y)) | i \in I, x, y \in V_i\}$ where on the right hand side of the equation, d is the distance in k^m.

In the sequel when we speak of a metric on an analytic manifold M it is of the metric associated with an admissible atlas on M.

3.9. Analytic Functions, Germs. A function $f : \Omega \to k$ on an open set Ω in an analytic manifold $(M, \underline{A})$ is *analytic* if the following holds: let $\mathcal{U} = \{(U_i, \Phi_i) | i \in I\}$ be an atlas in $\underline{A}$; then for every $i \in I$, $f \circ \Phi_i^{-1}$ is analytic on $\Phi_i(\Omega \cap U_i)$. It is obvious that this definition is independent of the choice of the atlas in $\underline{A}$. We denote the k-algebra of all analytic functions on Ω by $\mathcal{O}_{\underline{A}}(\Omega)$ ($\mathcal{O}_M(\Omega)$ or $C^\omega(\Omega)$ if $\underline{A}$ is known in the context). For a point $p \in M$ we denote by $\mathcal{O}_{Mp}$ the inductive limit of the $\mathcal{O}_M(U)$ as U varies over open neighbourhoods of p decreasing to p, the map $\mathcal{O}_M(U) \to \mathcal{O}_M(V)$ for $V \subset U$, being the restriction: $\mathcal{O}_{Mp}$ is the algebra of *germs* of k-valued analytic functions at p. From the definition one sees (using a chart (U, Φ) with $p \in U$) that $\mathcal{O}_{Mp} \simeq k\{\{X\}\}$, the algebra of convergent power series in m variables over k. The evaluation map $f \rightsquigarrow f(p)$, $f \in \mathcal{O}_M(U), p \in U$ of $\mathcal{O}(U)$ in k defines in the limit an algebra homomorphism $ev_p : \mathcal{O}_{Mp} \to k$ whose kernel $\mathfrak{m}_p$ is the unique maximal ideal in $\mathcal{O}_{Mp}$; $\mathfrak{m}_p$ maps isomorphically onto $\mathfrak{m}_o \subset k\{\{X\}\}$ (see 2.30 (3)). We remark here that when k is archimedean, the principle of analytic continuation holds: if an analytic function f on a connected open subset Ω in an analytic manifold vanishes on a non-empty open subset Ω' of Ω, it vanishes on all of Ω.

3.10. Examples. We now give some examples of analytic manifolds:

(**1**) An open set $U \subset k^m$ has a natural structure of an analytic manifold: an atlas is given by $\{(U, i : U \subset k^m)\}$. Evidently $\mathcal{O}_{k^m}(U)$ for an open set $U \subset k^n$ is the algebra of all k-valued analytic functions on U. The restriction of an analytic function, in particular, of a polynomial on k^n, to U is evidently an analytic function. Thus there are plenty of analytic functions on U.

(**2**) If $(M, \underline{A})$ is an analytic manifold and $\Omega \subset M$ is open, Ω inherits an analytic structure: an atlas for the analytic structure on Ω is given by $\{(U_i \cap \Omega, \Phi_i|_{U_i \cap \omega})| \; i \in I\}$ where $\{(U_i, \Phi_i)|i \in I\}$ is an atlas in $\underline{A}$. Evidently $\mathcal{O}_\Omega(U) = \mathcal{O}_M(U)$ for any open subset U of Ω. From (1) above we see that if (U, Φ) is a chart and $V \subset U$ is an open set, there are plenty of analytic functions on V.

(**3**) Let $U \subset k^{(l+m)}$ be open and $F = \{f_i| \; 1 \leq i \leq l\} : U \to k^l$ be an analytic function. Assume that $rank \; \{(\partial f_i/\partial x_j)(x)\}_{1 \leq i \leq l, 1 \leq j \leq (l+m)} = l$ for all $x \in U$. Let $p \in k^l$ and $Z = \{x \in U|F(x) = p\}$. Then Z has a natural structure of an analytic manifold of dimension m. A function f on an open set Ω of Z is analytic if and only if there is an open set V in U containing Ω and an analytic function $\tilde{f}$ on V which restricts to f on Ω (see 2.48 (Implicit Function Theorem)). Thus we see that in this case too there are many analytic functions.

(**4**) $\mathbb{S}^m = \{x = \{x_i\}_{1 \leq i \leq (m+1)} \in k^{(m+1)}| \; \sum_{i=1}^{(m+1)} x_i^2 = 1\}$, the '$m$-dimensional sphere' has a structure of an analytic manifold of dimension m: this is a special case of (3).

(**5**) Suppose $(M, \underline{A})$ and $(N, \underline{B})$ are analytic manifolds of dimensions m and n respectively. Then there is a natural structure of $(m + n)$ dimensional analytic manifold on $M \times N$, the product space. If $\mathcal{U} = \{(U_i, \Phi_i)|i \in I\} \in \underline{A}$ and $\mathcal{V} = \{(V_j, \Psi_j)|j \in J\} \in \underline{B}$, then $\mathcal{U} \times \mathcal{V} = \{(U_i \times V_j, \Phi_i \times \Psi_j)|(i, j) \in I \times J\}$ is an analytic atlas on $M \times N$.

(**6**) The *m-dimensional projective space* $\mathbb{P}^m(k)$ over k (the set of all 1-dimensional vector subspaces of $k^{(m+1)}$) has a natural structure of an analytic m-manifold. For $1 \leq r \leq (m+1)$ let $i_r : k^m \to \mathbb{P}^m(k)$ be the inclusion defined as follows: for $x = (x_1, x_2, \cdots, x_m)$,

$$i_r(x) = k \cdot (x_1, x_2, \cdots, x_{(r-1)}, 1, x_r, x_{(r+1)}, \cdots, x_m).$$

$\mathbb{P}^m(k)$ is equipped with the topology in which a set $U \subset \mathbb{P}^m(k)$ is open if and only if $i_r^{-1}(U)$ is open in k^m for $1 \leq r \leq (m+1)$. Set $i_r(k^m) = U_r$ and $\Phi_r = i_r^{-1} : U_r \to k^m$. Then one has for $1 \leq r \neq s \leq (m+1)$, $\Phi_r(U_r \cap U_s) = \{x = \{x_t\}_{1 \leq t \leq m} \in k^m \mid x_s \neq 0\}$ and $\Phi_s \circ \Phi_r^{-1}(\{x_i\}_{1 \leq i \leq m}) = \{y_i\}_{1 \leq i \leq m}$ where $y_i = x_i/x_s$ for $i \neq (r-1)$ or r according as $r > s$ or $r < s$, and $y_{(r-1)} = 1/x_s$ if $r > s$ while $y_r = 1/x_s$ if $r < s$. It is then clear that $\{(U_r, \Phi_r) \mid 1 \leq r \leq (m+1)\}$ is an analytic atlas on $\mathbb{P}^m(k)$. Thus the m-dimensional projective space over k carries a natural structure of an analytic manifold over k.

3.11. Remarks. (**1**) When k is non-archimedean, given any finite subset S of an analytic manifold M and a k-valued function f on S, there is an analytic function $\tilde{f}$ on M such that $\tilde{f}|_S = f$. This is an immediate consequence of Theorem 3.4.

(**2**) The statement analogous to that in (1) is false if $k = \mathbb{C}$. It is true in the case $k = \mathbb{R}$ but is a somewhat deep result. The proof in fact makes use of deep theorems about complex manifolds.

(**3**) When $k = \mathbb{C}$ the situation is much more complicated. There is a class of complex manifolds known as Stein manifolds for which the anologue of the statement in (1) holds. In general there may be no non-constant analytic function on a complex manifold. Compact complex manifolds ($\mathbb{P}^n(\mathbb{C})$ among them) provide such examples.

3.12. Analytic Maps. A map $F : M \to N$ from an analytic manifold M (of dimension m) to an analytic manifold N (of dimension n) is *analytic* if the following holds: let (U, Φ) (resp. (V, Ψ)) be a coordinate chart on M (resp. N); then the map $\Psi \circ F \circ \Phi^{-1} : \Phi(U \cap F^{-1}(V)) \to \Psi(V)(\subset k^n)$ is analytic (note that $\Phi(U \cap F^{-1}(V))$ is an open set in k^m. Observe that an analytic map is necessarily continuous.
A map $F : M \to N$ is an *(analytic) diffeomorphism* if it is an analytic bijection such that its inverse $F^{-1} : N \to M$ is also analytic.
It is clear from the above definition that analyticity of a map between manifolds is a *local* property. In other words we have the following.

3.13. Lemma. *A map $f : M \to N$ is analytic if and only if for every $x \in M$ there is an open set $U \ni x$ such that $f|_U$ is analytic.*

It now follows from 2.32 that we have.

3.14. Lemma. *Let L, M, N be analytic manifolds and $f : L \to M, g : M \to N$ analytic maps. Then $g \circ f$ is analytic.*

3.15. Remarks. If M, N are analytic manifolds, the Cartesian projections $\pi_M : M \times N \to M$ and $\pi_N : M \times N \to N$ are analytic. Also a map $f \times g : V \to M \times N$ of an analytic manifold V in to $M \times N$ is analytic if and only if $f : V \to M$ and $g : V \to N$ are analytic. In particular, for $p \in M$ (resp. $q \in N$) the inclusion $j_p : N \to M \times N$ (resp. $i_q : M \to M \times N$), defined by setting $j_p(y) = (p, y)$ for $y \in N$ (resp. $i_q(x) = (x, q)$ for $x \in M$) is analytic.

3.16. The Tangent Space. Let M be an analytic manifold and $p \in M$. A *tangent vector L to M at p* is a k-linear form $L : \mathcal{O}_{Mp} \to k$ (see 3.6) satisfying the following condition: for f, g in $\mathcal{O}_{Mp}$,

$$L(f \cdot g) = L(f) \cdot g(p) + f(p) \cdot L(g). \tag{$*$}$$

The set of tangent vectors $T_p M$ forms a k-vector space under addition and multiplication by elements of k: it is the *tangent space to M at p*. If (U, Φ) is a coordinate chart at p then for $0 \leq i \leq m$, $f \rightsquigarrow (\partial(f \circ \Phi^{-1})/\partial x_i)(\Phi(p))$, $f \in \mathcal{O}_{Mp}$ are all evidently tangent vectors to M at p. From 2.24 (see also 2.2, 2.3) we conclude that these tangent vectors which we denote $\partial/\partial x_i|_p$ constitute a basis of $T_p M$. From 2.24, it follows also that $T_p M$ is the dual of the finite dimensional vector space $T_p^* M = \mathfrak{m}_p / \mathfrak{m}_p^2$, the *space of differentials at p*. For an analytic f on an open subset U of M and a point $p \in U$, df_p is defined as the image of the germ of $f - f(p)$ at p in $T_p^* M$.

3.17. Remark. Let k' be a subfield of k containing $\hat{\mathbb{Q}}$. Then there is a natural isomorphism of $T_p M$ (resp. $T_p^* M$) regarded as a k'-vector space on $T_p[R_{k/k'} M]$ (resp. $T_p^*[R_{k/k'} M]$).

3.18. Differential of a Map. Let $F : M \to N$ be an analytic map of the analytic manifold M to the analytic manifold N. Let $p \in M$ and $F(p) = q$. Then composing analytic functions in an open neighbourhood V of q with F to obtain analytic functions in a neighbourhood of p, one gets a k-algebra homomorphism $F^* : \mathcal{O}_{Nq} \to \mathcal{O}_{Mp}$. If $L : \mathcal{O}_{Mp} \to k$ is a tangent vector to M at p, evidently $dF_p(L) \overset{\text{def}}{=} L \circ F^*$ is a tangent vector to N at q. As F^* is k-linear we see that $dF_p : T_p M \to T_q N$ is linear: it is the *differential of F at p*. The proofs of the following two lemmas are left to the reader.

3.19. Lemma. *Let M, N be analytic manifolds and Ω an open subset in M. Let $F : \Omega \to N$ be an analytic map. Let $p \in \Omega$ and $q = F(p)$. Let (U, Φ) (resp. (V, Ψ)) be a coordinate chart on M (resp. N) with $p \in U$ (resp. $q \in V$). Let $\Psi \circ F \circ \Phi^{-1} = \{f_i | 1 \leq i \leq n\}$, a map of the open set $\Phi(U \cap F^{-1}(V))$ into k^n.*

Then the matrix $J_{\Phi,\Psi}(F,p)$ of dF_p referred to the bases $\{\partial/\partial x_j|_p| \ 1 \leq i \leq m\}$ and $\{\partial/\partial y_j|_q| \ 1 \leq j \leq n\}$ is $\{(\partial f_i/\partial x_j)(p)\}_{1 \leq i \leq n, 1 \leq j \leq m}$.

3.20. Lemma. *Let L, M, N be analytic manifolds and $F : L \to M$ and $G : M \to N$ analytic maps. Then for any $p \in L$, $d(G{\circ}F)_p = dG_{F(p)}{\circ}dF_p$. Moreover if $(U, A), (V, B), (W, C)$ are charts on L, M, N respectively with $p \in U$, $F(p) \in V$, $G(F(p)) \in W$, then $J_{A,C}(F{\circ}G, p) = J_{B,C}(G, F(p)) \cdot J_{A,B}(F, p)$. In other words the composition of the linear maps $dG_{F(p)}$ and dF_p corresponds to multiplication of the corresponding matrices.*

3.21. Definitions. Let M, N be analytic manifolds and $F : M \to N$ an analytic map.
F is an *immersion* if dF_p is an injection for every $p \in M$ (this implies dimension $M \leq$ dimension N). F is a *submersion* if dF_p is a surjection for every $p \in M$ (this implies dimension $M \geq$ dimension N). F is a *diffeomorphism* if F is a bijection and F^{-1} is analytic. Note that in this case $dF^{-1}_{f(p)}{\circ}dF_p$ is the identity for all $p \in M$ so that F is both an immersion and a submersion and dimension $M =$ dimension N.

Let M (resp. N) be an analytic manifold of dimension m (resp. n) and $F : M \to N$ an analytic map. Then we have the following reformulations of the inverse function theorem, the implicit function theorem and the rank theorem.

3.22. Inverse Function Theorem. *Suppose that $p \in M$ is such that dF_p is an isomorphism. Then there are open sets $U \ni p$ in M and $V \ni q = f(p)$ of N such that $F(U) = V$ and $F|_U : U \to V$ is a diffeomorphism.*

3.23. Implicit Function Theorem. *Suppose that $p \in M$ is such that dF_p is a surjection. Then there are charts (U, Φ) on M and (V, Ψ) on N with $p \in U$ such that $F(U) \subset V$, $F|_U$ is a submersion and $\Psi{\circ}F{\circ}\Phi^{-1}(x_1, x_2, \cdots, x_m) = (x_{m-n+1}, x_{m-n+2}, \cdots, x_m)$.*

3.24. Rank Theorem. *Suppose that rank $(dF_x) = r$ for all $x \in M$. Then for any $p \in M$ there are charts (U, Φ) on M and (V, Ψ) on N with $p \in U$ such that $F(U) \subset V$ and $\Psi{\circ}F{\circ}\Phi^{-1}(x_1, x_2, \cdots, x_n) = (x_1, x_2, \cdots, x_r, 0, \cdots, 0)$.*

3.25. An analytic *sub-manifold* of an analytic manifold M is a pair $(N, u : N \to M)$ where N is an analytic manifold and i is an *injective* (analytic) immersion. When u is known from the context, we denote (N, u) by simply N. Also we often identify N with its image in M, always exercising care about the possibility of the topology of N not being induced from M (see remark below).

3.26. Remarks. In general if (V, i) is an analytic sub-manifold of M, the topology induced on V from M through i is not the manifold topology on V. Three examples to illustrate this are given below, the first two in the case $k = \mathbb{R}$ and the third in the case of any k.

(1) Consider the analytic manifold $\mathbb{S}^1 \times \mathbb{S}^1$. Fix $\alpha \in \mathbb{R}$ and define $u : \mathbb{R} \to \mathbb{S}^1 \times \mathbb{S}^1$ by setting for $t \in \mathbb{R}$, $u(t) = (\exp(i \cdot t), \exp(i \cdot \alpha \cdot t))$; then one sees that $(\mathbb{R}, u)$ is an analytic sub-manifold if and only if α is irrational; and in that case the topology induced on $\mathbb{R}$ through u is strictly coarser than the standard topology on $\mathbb{R}$. It is easy to see that $u(\mathbb{R})$ is dense in $\mathbb{S}^1 \times \mathbb{S}^1$ (this follows from the fact for irrational $\alpha \in \mathbb{R}$, the set $\{a + b \cdot \alpha | \ a, b \in \mathbb{Z}\}$ is dense in $\mathbb{R}$).

(2) Let $V = (-\pi/4, \pi/8) \subset \mathbb{R}$ and $i : V \to \mathbb{R}^2$ be the analytic map defined by setting $i(t) = (\mathrm{Cos}(t) \cdot \mathrm{Cos}(4 \cdot t), \mathrm{Sin}(t) \cdot \mathrm{Cos}(4 \cdot t))$; then, as is easily checked, (V, i) is a sub-manifold of $\mathbb{R}^2$. In the topology on $(-\pi/4, \pi/8)$ induced by i, $\{(-\pi/8 - \delta, -\pi/8 + \delta) \cup (\pi/8 - \delta, \pi/8) | \ 0 < \delta < \pi/8\}$ is a fundamental system of neighbourhoods of $-\pi/8$.

(3) Let $\{a_n\}_{n \in \mathbb{N}}$ be a sequence in k such that $a_0 = 0$, $a_n \neq 0$ for $n > 0$ and a_n tends to 0 as n tends to infinity. Let V be the 1-dimensional manifold $k \times \mathbb{N}$ and $i : k \times \mathbb{N} \to k^2$, the map defined by setting $i(x, n) = (x, a_n)$. A fundamental system of neighbourhood of $(0, 0) \in k \times \mathbb{N}$ in the topology induced by i is given by $\{(x, a_n) | x \in k, |x| < \delta, |a_n| < \delta\}$.

3.27. However in many naturally occurring cases, the topology induced from u is the same as the underlying topology of V as an analytic manifold. This is notably the case of the inclusion of $F^{-1}(q)$ in M where $F : M \to N$ is an analytic map of analytic manifolds and $q \in N$ is such that dF_p is surjective for all $p \in F^{-1}(q)$ (this is an immediate consequence of the implicit function theorem).

Next suppose that $(V, i : V \to M)$ is a sub-manifold such that $i : V \to M$ is a proper map (i.e., $i^{-1}(K)$ is compact for every compact subset K of M). Denote by N' the topological space $i(N)$ with the topology induced from M. Let $S = \{y_n | \ n \in \mathbb{N}\}$ be a sequence in N' converging to y. Then $S_o = S \cup \{y\}$ is a compact subset M. It follows that $i^{-1}(S_o)$ is *compact* and as i is injective $i^{-1}(S_o) = i^{-1}(S) \cup \{x\}$ for a (unique) $x \in N$ with $i(x) = y$. It follows that the sequence $i^{-1}(S)$ converges to $x = i^{-1}(y)$ proving that $i^{-1} : N' \to N$ is continuous. We have thus proved the first assertion of the following proposition. We leave the second assertion as an exercise to the reader.

3.28. Proposition. *Let $(V, i : V \to M)$ be an analytic sub-manifold of an analytic manifold M. If i is a proper map, the topology on V is the same* *as that induced from M by i. For every $p \in V$, there is a coordinate chart (U, Φ) at $i(p)$ (on M) with $\Phi(p) = 0$ and a linear subspace L of k^m such that $i(V) \cap U = \Phi^{-1}(L)$.*

The following proposition is useful when dealing with sub-manifolds whose manifold topology is not induced from the ambient manifold.

3.29. Proposition. *Let $(N, i : N \to M)$ be an analytic sub-manifold of the analytic manifold M. Then a map $f : P \to N$ of an analytic manifold P to N is analytic iff it is continuous and $i \circ f$ is analytic.*

3.30. Proof. Let $x \in N$ and $(U, F : U \to \Omega \subset k^m)$ be a coordinate chart on M at $i(f(x))$ such that $i^{-1}(\{z \in U|\ z_i = 0\ for\ i > n\})$ ($n = \dim N$) is an open neighbourhood V of $f(x)$ in N. Also if p is the Cartesian projection on the first n coordinates, $(V, G = p \circ i : V \to \mathbb{R}^n)$ is a chart on N at $f(x)$. Since f is continuous $f^{-1}(V) = W$ is open and $G \circ f = p \circ (F \circ i)$ is analytic we see that f is analytic. The implication in the other direction is obvious.

3.31. Vector fields. $A(n\ analytic)$ *vector field* X on an open set U_o in an analytic manifold M is, an assignment to each $p \in U_o$, of a tangent vector $X(p)$ (also denoted X_p) in T_pM such that the following holds: let (U, Φ) be a coordinate chart on M; then one has for every $q \in U$, a basis $\{\partial/\partial x_i|_q|\ 1 \le i \le m\}$ of T_qM; thus for $p \in U_o \cap U$, there are uniquely determined $\{a_i(p) \in k|\ 1 \le i \le m\}$ such that $X(p) = \sum_{i=1}^m a_i(p) \cdot \partial/\partial x_i$; then the a_i are analytic functions on $U_o \cap U$.

If (U, Φ) is chart with $U \subset U_o$, then the assignments $p \leadsto \partial/\partial x_i|_p, p \in U, 1 \le i \le m$, are evidently analytic vector fields on U. If X is a vector field on an open set U of M and $f \in \mathcal{O}(U)$, we denote by $X \cdot f$ the function in $\mathcal{O}(U)$ defined by setting for $p \in U$, $(X \cdot f)(p) = X(p)(f)$; it is clear from the definitions that $X \cdot f$ is an analytic function on U. Thus we have a k-linear endomorphism $X : \mathcal{O}(U) \to \mathcal{O}(U)$. This endomorphism evidently satisfies the following condition: for $f, g \in \mathcal{O}(U)$,

$$X \cdot (f \cdot g) = (X \cdot f) \cdot g + f \cdot (X \cdot g).$$

3.32. If X is an analytic vector field on U and $f \in \mathcal{O}(U)$, the assignment $p \leadsto f(p) \cdot X(p)$ is evidently an analytic vector field which is denoted $f \cdot X$. If X, Y are vector fields on U, we define $X + Y$ as the vector field $(X + Y)(p) = X(p) + Y(p)$, $p \in U$. We denote by $\mathcal{T}(U)$ the set of all vector fields on U; then the addition and the multiplication by functions defined above make $\mathcal{T}(U)$ into a module over $\mathcal{O}(U)$. For open sets U, V in M, $X \in \mathcal{T}(U)$ and $f \in \mathcal{O}(U)$ one has evidently $(X \cdot f)|_V = X|_V \cdot f|_V$. It follows that for any $p \in U$, X defines an endomorphism of $\mathcal{O}_{Mp}$. Thus if X and Y are in $\mathcal{T}(U)$ we have endomorphisms $X \circ Y$ which we denote $X \cdot Y$. We define $[X, Y]$ as the endomorphism $X \cdot Y - Y \cdot X$ of $\mathcal{O}_{Mp}$.

3.33. Proposition. *If $ev_p : \mathcal{O}_{Mp} \to k$ is the evaluation at p of the germs of analytic functions at p, then $[X, Y](p) \overset{\text{def}}{=} ev_p \circ [X, Y] : \mathcal{O}_{Mp} \to k$ is a tangent vector at p. The assignment $[X, Y](p)$ to every $p \in U$ defines an analytic vector field on U.*

$[X, Y]$ above is the *bracket* or *Lie bracket* of X and Y.

3.34. Proof. If V, W are open sets in U, with $p \in V \subset W$, then for $X, Y \in \mathcal{T}(W)$, $[X, Y]|_V = [X|_V, Y|_V]$ as is easily checked. One has for analytic functions f, f' on V,

$$[X, Y](f \cdot f') = X(Y(f \cdot f')) - Y(X(f \cdot f')) =$$
$$X(Yf \cdot f' + f \cdot Yf') - Y(Xf \cdot f' + f \cdot Xf') =$$
$$(X(Yf) \cdot f' + Yf \cdot Xf' + Xf \cdot Yf' + f \cdot X(Yf')) - (YXf \cdot f' +$$
$$Xf \cdot Yf' + Yf \cdot Xf' + f \cdot YXf').$$

It follows that $[X, Y](f \cdot f') = [X, Y]f \cdot f' + f \cdot [X, Y]f'$. It is now immediate that $[X, Y](p)$ is a tangent vector. In a local chart any analytic vector field is of the form $\sum_{i=0}^{m} a_i \cdot \partial/\partial x_i$ with a_i analytic. The analyticity of $[X, Y]$ thus follows from that of $[a \cdot (\partial/\partial x_i), b \cdot (\partial/\partial x_j)]$ for analytic functions a and b and integers $i, j \leq m$. This last bracket is in fact equal to $a \cdot \partial b/\partial x_i \cdot \partial/\partial x_j - \beta \cdot \partial a/\partial x_i \cdot \partial/\partial x_j$, an analytic vector field. This proves the proposition.

3.35. Proposition. *Let U be an open set in M and X, Y vector fields in U. Then $\mathcal{T}(U)$ is a Lie algebra over k under the bracket operation $(X, Y) \rightsquigarrow [X, Y]$ defined above.*

Evidently $[X, X] = 0$ for all $X \in \mathcal{T}(U)$. So we need to only check that for $X, Y, Z \in \mathcal{T}(U)$ $[X, [Y, Z]] + [Y, [Z, X]] + [Z, [X, Y]] = 0$ which is straightforward. Note that the bracket operation is <u>not</u> bilinear over $\mathcal{O}(U)$, so $\mathcal{T}(U)$ is not a Lie algebra over $\mathcal{O}(U)$.

3.36. Definition. *By an (analytic) family of (analytic) vector fields on a manifold M parametrized by a manifold Ω we mean a collection $\{X(\alpha) | \alpha \in \Omega\}$ of vector fields $X(\alpha)$ on M satisfying the following condition: for charts (U, Φ) on M and (A, Γ) on Ω, and $\alpha \in A$, let $X(\alpha)|_U = \sum_{i=1}^{m} f_i(\alpha, x)(\partial/\partial x_i)$; then $f_i : A \times U \to k$ are analytic functions.*

With this definition we have the following.

3.37. Theorem. *Let $\{X(a) | a \in \Omega\}$ be an analytic family of analytic vector fields on M. Then given compact subsets A in M and Γ in Ω, there is a $\delta > 0$, open sets U of M and Δ of Ω with $A \subset U$, $\Gamma \subset \Delta$, and an analytic map $\Phi : \Delta \times U \times I_\delta(0) \to M$ satisfying the following conditions:*
(i) $\Phi(a, x, 0) = x$ for all $x \in U$ and $a \in \Delta$.
(ii) $\Phi(a, \Phi(a, x, s), t) = \Phi(a, x, t + s)$ for all $a \in \Delta$, $t, s \in I_\delta(0)$ and $x \in U$ with $(t + s) \in I_\delta(0)$ and $\Phi(a, x, t) \in U$.

(iii) *For any analytic function u defined on an open subset $W \subset U$ with $x \in W$ and $t \in I_\delta(0)$,*

$$(X(a) \cdot u)(\Phi(a, x, t)) = d(u(\Phi(a, x, t)))/dt.$$

3.38. Proof. We first take up the case when A and Γ are singleton sets. Let $(a, p) \in \Gamma \times A$ be any point. Let (U, F) (resp. (V, G)) be a chart in M (resp. Ω) with $p \in U$ (resp. $a \in V$). We will now show that there is an open neighbourhood U' (resp. V') of p (resp. a) contained in U (resp. V), a $\delta_p > 0$ and an analytic map $\Phi_{U', V'} : U' \times V' \times I_{\delta_p} \to U$ satisfying (i)–(iii) above with $\Phi_{U', V'}$, U', V' and δ_p replacing Φ, M, Ω, Δ and δ respectively. Now in U', the vector field $X(a)$ can be written as $\sum_{i=0}^{m} f_{ia} \cdot (\partial/\partial x_i)$ where $f_{ia} = X(a)(x_i)$; it follows that if u is an analytic function on an open subset of U, we have $X(a) \cdot u = \sum_{i=0}^{m} f_{ia} \cdot (\partial u/\partial x_i)$. Treating Φ as well as the vector field as k^n-valued functions, we see that the existence of Φ is equivalent to the existence and uniqueness of a solution of the following system of ordinary differential equations parametrized by V:

$$d\Phi(a, \xi, t)/dt = F(\Phi(a, \xi, t))$$

with the initial condition

$$\Phi(a, \xi, 0) = \xi.$$

Note that $F = \{f_i\}_{1 \le i \le m}$ is independent of t and the parameter (in V'); and Theorem 2.54 guarantees the existence of Φ. Condition (ii) follows from the uniqueness of the germ of the solution.

The case of a general $A \times \Gamma$ follows by arguments similar to those in 2.54.

3.39. Corollary. *Let X be a vector field on M and K a compact subset of M. Then for any relatively compact open subset of U of M containing K, there is a $\delta = \delta(U) > 0$ and an analytic map $\Phi_X : U \times I_\delta(0) \to M$ satisfying the following conditions:*
(i) $\Phi_X(x, 0) = x$ *for all $x \in U$.*
(ii) $\Phi_X(\Phi_X(x, s), t) = \Phi_X(x, t + s)$ *for all $t, s \in I_\delta(0)$ with $(t + s) \in I_\delta(0)$ and $x \in U$ with $\Phi_X(x, s) \in U$.*
(iii) *For an analytic function u defined on an open set $V \subset U$ and $x \in V$ and $t \in I_\delta(0)$,*

$$d(u(\Phi_X(x, t)))/dt = (X \cdot u)(\Phi_X(x, t)).$$

3.40. Remarks. (1) For a vector field X on M and $t \in I_\delta(0)$ we denote by $\varphi_{X,t}$ or simply φ_t (when the vector field being considered is known from the context), the analytic map $U \ni x \rightsquigarrow \Phi(x, t)$. With this notation we have

$$(\varphi_t \circ \varphi_s)(x) = \varphi_{(t+s)}(x) \tag{$*$}$$

for $t, s \in I_\delta(0)$ and $x \in U'$ when both sides are defined. $\{\varphi_t\}_{t \in I_\delta(0)}$ is the 'local 1-parameter group of local diffeomorphisms' of the vector field X. If $U = M$ is compact, for $t \in I_\delta(0)$, φ_t is defined on all of M.

(**2**) When k is non-archimedean and M compact $I_\delta(0)$ is an open and closed subgroup of the additive group k and $t \rightsquigarrow \varphi_t$ is a homomorphism of the $I_\delta(0)$ into the group of all analytic automorphisms of M.

(**3**) When k is archimedean $I_\delta(0)$ is not a subgroup. However in the case when M is compact (or more generally if φ_t is defined on all of M for all $t \in I_\delta(0)$ for some $\delta > 0$) then φ_t is defined for *all* $t \in k(= \mathbb{R} \text{ or } \mathbb{C})$ and the map $t \rightsquigarrow \varphi_t$ is a homomorphism of k into the group of all analytic automorphisms of M. This is seen as follows: let $\delta' > 0$ be such that φ_t is defined for $|t| < \delta'$ and set $\delta = \delta'/2$. Then given any element $z \in \mathbb{R}$ (resp. $\mathbb{C}$), there is a unique integer $a(z)$ (resp. pair of integers $(a(z), b(z))$) such that $0 \leq (z - a(z) \cdot \delta) < \delta/2$ (resp. $(z - (a(z) \cdot d + i \cdot b(z) \cot \delta)) = u + i \cdot v$ with $0 \leq u < \delta/2, 0 \leq v < \delta/2)$. One then defines for $z \in \mathbb{R}$ (resp. $\mathbb{C}$), $\varphi_z = \varphi_\delta^{a(z)} \circ \varphi_{(z-a(z))}$ (resp. $\varphi_z = \varphi_\delta^a \circ \varphi_{i \cdot \delta}^{b(z)} \circ \varphi_{(z-(a(z)+i \cdot b(z)))})$. It is then easily seen that $t \rightsquigarrow \varphi_t$ is a homomorphism of (the additive group) $k \simeq \mathbb{R}$ or $\mathbb{C}$ in the group of analytic diffeomorphisms of M on itself.

(**4**) Let k' be a locally compact subfield of k. Then there is a natural identification of the tangent space to M (over k) at a point p with the tangent space to M as an analytic manifold over k' and hence every vector field on M over k defines a vector field on M over k' as well. It follows easily from the definitions that the local 1-parameter group for a vector field X as a vector field over k' is the restriction to $I_\delta \cap k'$ of the 1-parameter group of X as a vector field over k.

3.41. Lemma. *Let X be a vector field in an open set Ω of an analytic manifold M. If $p \in \Omega$ is such that $X(p) \neq 0$, then there is a coordinate chart (V, Ψ) with $p \in V \subset U$ such that $X|_V = \partial/\partial x_1$ in the chart.*

3.42. Proof. Fix a chart (U, Ψ) with $p \in U \subset \Omega$ and $\Psi(p) = 0$. We identify U with an open set in k^m via Ψ. Composing Ψ with a linear map $L : k^m \to k^m$, we assume that $X(p) = \partial/\partial x_1|_p$. Then $X = \sum_{i=0}^m f_i \cdot \partial/\partial x_i$ in U with $f_i(0) = \delta_{1i}$. Let U' be an open set with $0 \in U' \subset U$ and $\delta > 0$ be such that there is a $\Phi : U' \times I_\delta(0) \to U$ as in Proposition 3.28. Let $W = \{x \in U'|x_1 = 0\}$. Define the analytic map $F : W \times I_\delta(0) \to U$ by setting for $(w, t) \in W \times I_\delta(0)$, $F(w, t) = \Phi(w, t)$; then (under the natural identifications of the tangent space at any point of U' or $W \times I_\delta(0)$ with k^m), $dF_{0,0}$ is the identity. It follows that there is an open set W' in W and $\delta > 0$ such that $F|_{W' \times I_\delta(0)}$ is an analytic diffeomorphism on to an open subset $V \ni 0$ of U' (Inverse Function Theorem). It is easily seen that in the coordinate system $(\Psi^{-1}(V), F \circ \Psi)$ the vector field takes the form $\partial/\partial x_1$. Hence the lemma.

The proof of the following proposition is easy and left to the reader.

3.43. Proposition.   *The 1-parameter group* $\varphi_{(\partial/\partial x_j)t}$ *of* $\partial/\partial x_j$ *in* k^n *is given by* $\varphi_{(\partial/\partial x_j)t}\left(\{x_i\}_{1\leq i\leq m}\right) = \{x_i + t\cdot\delta_{i,j}\}_{1\leq i\leq m}$.

3.44. Let X, Y be analytic vector fields on an open set U of the analytic manifold M and W an open neighbourhood of a point $p\in U$ with compact closure $\bar{W}$. Let $\{\varphi_t : W\to U|\ |t| < \delta\}$, $\delta > 0$ be the local 1-parameter group of local diffeomorphisms determined by X. Evidently, if W' is a relatively compact open neighbourhood of p with closure contained in W, there is a positive $\delta' < \delta$ such that $\varphi_t(W')\subset W$ for $|t| < \delta'$. For $t\in k$ with $|t| < \delta'$ and $q\in W'$ set $\varphi_{t*}(Y)(q) = d(\varphi_t)_{\varphi_{-t}(q)}(Y(\varphi_{-t}(q)))$. Then it is easily seen that for each $t\in I_{\delta'}$, $\varphi_{t*}(Y)(q)$ is an analytic vector field on W' and for fixed q, $\varphi_{t*}(Y)(q)$ is an analytic function on $I_{\delta'}$ with values in the tangent space T_qM to M at q. Set $Z(q) = \frac{d}{dt}(\varphi_{t*}(Y)(q))|_{t=0}$. Then $q\rightsquigarrow Z(q)$ is analytic vector field on all of U (we vary the point q over all of U). We then have the following.

3.45. Proposition. $Z = [X,Y]$.

3.46. Proof. Let $U' = \{q\in U|X(q)\neq 0\}$. Then by the Lemma above for any $q\in U'$, there is a chart (V,Ψ) with $q\in V\subset U'$ such that $X = \partial/\partial x_1$. It follows that in V, $\varphi_t(x_1, x_2, \cdots, x_m) = (x_1 + t, x_2, \cdots, x_m)$ (see (1) of 3.26). In the chart let $Y = \sum_{1\leq i\leq m} a_i(x)\cdot\partial/\partial x_i$. It is immediate from this that in V, $Z = [X,Y]$; and since $q\in U'$ is arbitrary, we see that $Z = [X,Y]$ on all of U'. Let U'' be the set of points p in U such that $X = 0$ in a neighbourhood of p in U. U'' is open in U so that $[X,Y] = 0$ in U''. On the other hand for $p\in U''$ the local 1-parameter group φ_t is the identity map for all t near 0 so that $Z(p) = \frac{d}{dt}(\varphi_{t*}(Y)(p))|_{t=0} = 0$. Thus $Z = [X,Y]$ on $U'\cup U''$. As both Z and $[X,Y]$ are analytic and $U'\cup U''$ is dense in U (U'' is the complement of the closure of U' in U), $Z = [X,Y]$ on all of U.

3.47. Corollary. *Let* X, Y *be vector fields on an open set* $U\subset M$. *Then if* $[X,Y] = 0$, $\varphi_{Xt}\circ\varphi_{Ys} = \varphi_{Ys}\circ\varphi_{Xt}$ *on a neighbourhood of any point* $p\in U$ *and for all* t, s *sufficiently near 0 for which both the composite maps are defined, and conversely.*

3.48. Proof. The local 1-parameter group of the vector field $\varphi_{Xt*}(Y)$ (in a suitable neighbourhood of any given point) is $\varphi_{Xt}\circ\varphi_{Ys}\circ\varphi_{X(-t)}$ – this follows from the uniqueness of the solution in the initial value problem for ordinary differential equations. Now if $[X,Y] = 0$, we see that $d/dt(\varphi_{Xt*}(Y)) = 0$ (Proposition 3.28) i.e., $\varphi_{Xt*}(Y)$ is constant and hence equals Y. It follows that $\varphi_{Xt}\circ\varphi_{Ys}\circ\varphi_{X(-t)}$ is the local 1-parameter group of Y and thus is equal to φ_{Ys}, proving one part of the corollary; the converse is a trivial consequence of Proposition 3.28.

3.49. Differential Forms. Recall that for a vector space E of dimension m and an integer $p \geq 0$ over k, $\bigwedge^{p*} E$ denotes the space of all multilinear alternating p-forms on E (when $p = 0$, $\bigwedge^{0*} = k$) . Then $\dim_k \bigwedge^{p*} E = \binom{m}{p}$. In particular, if $p > m$, $\bigwedge^{p*} E = \{0\}$.

An (analytic) exterior differential p-form (or simply a *p-form*) α on an analytic manifold M is an assignment to each $z \in M$, of an element $\alpha_z \in \bigwedge^{p*} T_z M$ satisfying the following condition: if (U, Φ) is a coordinate chart and $\{i_1, i_2, \cdots, i_p\}$ is any p-tuple in $[1, m]$, the function

$$U \ni z \rightsquigarrow \alpha_z(\partial/\partial x_{i_1}|_z, \partial/\partial x_{i_2}|_z, \cdots, \partial x_{i_p}|_z)$$

is an analytic function on U. We denote vector space of all p-forms on M by $\Omega^p(M)$ and the direct sum $\coprod_{p=0}^m \Omega^p(M)$ by $\Omega(M)$. The lemma below follows from the fact that analyticity of a function is a local property.

3.50. Lemma. *Let α be an assignment $M \ni z \rightsquigarrow \alpha_z \in \bigwedge^{p*} T_z M$ of an alternating multilinear p-form on $T_z M$ to each $z \in M$. Then α is an analytic p-form on M if and only if for every $z \in M$, there is an open neighbourhood W_z of z such that $\alpha|_{W_z}$ is an analytic p-form.*

3.51. The Graded Algebra $\Omega(M)$. We now define a multiplication $\wedge$ on $\Omega(M)$ as follows: For $\alpha \in \Omega^p(M)$, $\beta \in \Omega^q(M)$, $z \in M$ and $v_1, v_2, \cdots, v_{p+q} \in T_z M$,

$$(\alpha \wedge \beta)_z(v_1, v_2, \cdots, v_{p+q}) =$$

$$\left\{ \sum_{\sigma \in S_{p+q}} \epsilon_\sigma \cdot \alpha_z(v_{\sigma(1)}, v_{\sigma(2)}, \cdots, v_{\sigma(p)}) \cdot \beta_z(v_{\sigma(p+1)}, v_{\sigma(p+2)}, \cdots, v_{\sigma(p+q)}) \right\}/r!$$

where $r = p + q$ and for $\sigma \in S_{p+q}$, ϵ_σ is the signature of σ. If $\alpha \in \Omega^0(M)$, it is an analytic function f and $\alpha \wedge \beta$ is defined as $f \cdot \beta$. For general elements $u, v \in \Omega(M)$ $u \wedge v$ is defined by distributivity. It is easily checked that the following lemma holds so that $\Omega(M)$ is an anti-commutative graded algebra over k with $\Omega^p(M)$ as the pth graded component.

3.52. Lemma. *For $\alpha, \beta, \gamma \in \Omega(M)$, $(\alpha \wedge \beta) \wedge \gamma = \alpha \wedge (\beta \wedge \gamma)$. For $\alpha \in \Omega^p(M)$ and $\beta \in \Omega^q(M)$, $\alpha \wedge \beta = (-1)^{p \cdot q} \cdot \beta \wedge \alpha$.*

3.53. Suppose now that $F : M \to N$ is an analytic map; then for $\alpha \in \Omega^p(M)$ one defines the form $F^*(\alpha)$ *induced by α through F* as follows: for $z \in M$ and tangent vectors $v_1, v_2, \cdots, v_p \in T_z M$,

$$F^*(\alpha)(v_1, v_2, \cdots, v_p) = \alpha(dF_z(v_1), dF_z(v_2), \cdots, dF_z(v_p)).$$

That the analyticity requirement is satisfied by $F^*(\alpha)$ follows from the fact that if (U, Φ) and (V, Ψ) are charts on M and N respectively with $F(U) \subset V$, the matrix of dF referred to the bases of $T_z M$ and $T_{F(z)} N$ given by the coordinate systems is the Jacobian matrix $\{\partial F_i/\partial x_j\}_{1 \leq i \leq n; 1 \leq j \leq m}$ where F_i

is the ith component of the (vector-valued) function $\Psi \circ F \circ \Phi^{-1}$. The proof of the following lemma is left as an exercise to the reader.

3.54. Lemma. (i) $F^* : \Omega(n) \to \Omega(M)$ *is a graded k-algebra morphism.*
(ii) *If $F : L \to M$ and $G : M \to N$ are analytic maps $(G \circ F)^* = F^* \circ G^*$.*

3.55. Let (V, Φ) be a chart and $\mathcal{I}(p)$ the set of all subsets of cardinality p in $[1, m]$. For $I = \{i_1 < i_2 < \cdots < i_p\} \in \mathcal{I}(p)$, let dx_I denote the p-form defined by setting for $z \in V$ and $J = \{j_1 < j_2 < \cdots < j_p\} \in \mathcal{I}(p)$,

$$dx_{Iz}(\partial/\partial x_{j_1}|_z, \partial/\partial x_{j_2}|_z, \cdots, \partial x_{j_p}|_z) = \delta_{I,J}.$$

The $\{dx_{Iz}|\ I \in \mathcal{I}(p)\}$ then constitute a basis for the space of all alternating p-forms on $T_z M$. It follows that any differential p-form α can be expressed uniquely as a linear combination

$$\sum_{I \in \mathcal{I}(p)} a_I \cdot dx_I$$

where a_I is the analytic function defined by setting for $z \in V$

$$a_I(z) = \alpha_z(\partial/\partial x_{i_1}|_z, \partial/\partial x_{i_2}|_z, \cdots, \partial x_{i_p}|_z).$$

3.56. Exterior Differentiation. For an open set $U \subset k^m$ define the k-linear map $d_U : \Omega^p(U) \to \Omega^{p+1}(U)$ as follows: let $\omega \in \Omega^p(U)$; then if $\omega = \sum_{I \in \mathcal{I}(p)} \omega_I \cdot dx_I$, we set

$$d_U(\omega) = \sum_{I \in \mathcal{I}(p)} \sum_{i \in [1,m]} (\partial \omega_I / \partial x_i) \cdot dx_i \wedge dx_I.$$

Note that $dx_i \wedge dx_I = 0$ if $i \in I$ and if $i \notin I$, $dx_i \wedge dx_I = (-1)^q \cdot dx_{\{i\} \cup I}$ where $q = q(i, I)$ is the unique integer such that $i_q < i < i_{q+1}$ (here we have set $i_0 = 0$ and $i_{p+1} = m + 1$).
If U, U' are open subsets of k^m and $F : U \to U'$ is an analytic map, then from the chain rule for differentiation, one deduces easily the following.

3.57. Theorem. *The diagram*

$$\begin{array}{ccc} \Omega^p(U) & \stackrel{d_U}{\to} & \Omega^{p+1}(U) \\ \downarrow F^* & & F^* \downarrow \\ \Omega^p(U') & \stackrel{d'_U}{\to} & \Omega^{p+1}(U') \end{array}$$

is commutative.

3.58. Suppose now that M is an analytic manifold. Then we define $d_M : \Omega^p(M) \to \Omega^{p+1}(M)$ as follows: for $\omega \in \Omega^p(M)$ and any coordinate chart (U, Φ) on M,

$$(d_M \omega)|_U = \Phi^*(d_{\Phi(U)}(\Phi^{-1*}(\omega)))$$

where Φ^{-1} is treated as a map of the open set $\Phi(U)$ into M. Since any point is contained in U for some chart, in view of Theorem 3.57 above, $d_M(\omega)$ is well defined. It follows from the theorem that we have

3.59. Corollary. *If $F : M \to N$ is an analytic map, the diagram*

$$\Omega^p(N) \overset{d_N}{\to} \Omega^{p+1}(N)$$
$$\downarrow F^* \qquad F^* \downarrow$$
$$\Omega^p(M) \overset{d_M}{\to} \Omega^{p+1}(M)$$

is commutative.

3.60. Remarks: Connection with Vector Analysis. (1) A p-form $\omega = \sum_{I \in \mathcal{I}(p)} \omega_I \cdot dx_I$ on an open set U in k^n can be thought of as the vector valued function $U \ni x \leadsto \{\omega_I\}_{I \in \mathcal{I}(p)}$ on U with values in $k^{\binom{n}{p}}$. When $n = 3$ we have $\binom{3}{0} = 1 = \binom{3}{3}$ and $\binom{3}{1} = 3 = \binom{3}{2}$. Thus 0- and 3-forms on U are identified with scalar functions while 1- and 2-forms are identified with vector (i.e., k^3) valued functions. Under these identifications, one sees that $\alpha \wedge \beta$ for a 0-form α and any β is multiplication by the scalar function. If α and β are 1-forms, they are vector valued functions and their product $\alpha \wedge \beta$ – a 2-form – is nothing but the cross-product. Finally the product of a 1-form and a 2-form treated as vector valued functions is their dot-product.

(2) Exterior differentiation too translates to familiar operations: d_U on 0-forms is the gradient (grad), on 1-forms is the curl and on 2-forms it is the divergence (div) of 'vector analysis'. One has $curl \circ grad = 0$ and $div \circ curl = 0$. These have the following generalization to general analytic manifolds.

3.61. Theorem. (i) *For $\alpha \in \Omega^p$ and $\beta \in \Omega^q$,*

$$d_M(\alpha \wedge \beta) = (d_M \alpha) \wedge \beta + (-1)^p \alpha \wedge (d_M \beta).$$

(ii) $d_M(d_M(\alpha)) = 0$ *for all $\alpha \in \Omega^*(M)$.*

3.62. Proof. Evidently it suffices to prove the theorem in the case when M is an open subset of k^n. For proving the first statement we need to only consider the case $\alpha = f \cdot dx_I$ and $\beta = g \cdot dx_J$ where f and g are analytic functions on U and $I \in \mathcal{I}(p)$ and $J \in \mathcal{I}(q)$ – linearity takes care of the general case. It is clear from the definitions that $d_M(dx_I) = 0$. It is also easy to see from the definitions that

$$d_M(f \cdot dx_I) = \sum_{i \in [1,n]} (\partial f / \partial x_i) \cdot dx_i \wedge dx_I \qquad (1)$$

$$d_M(f \cdot dx_I \wedge g \cdot dx_j) = d_M(f \cdot dx_I) \wedge g \cdot dx_j + (-1)^{c(I)} \cdot f \cdot dx_I \wedge d_M(g \cdot dx_j) \quad (2)$$

where $c(I)$ is the cardinality of I. Now $f \cdot dx_I \wedge g \cdot dx_J = (f \cdot g) \cdot (dx_I \wedge dx_J)$ as is clear from the definitions. It follows that

$$d_M((f \cdot g \cdot dx_I) \wedge dx_j) = \sum_{h \in [1,n]} (\partial (f \cdot g) / \partial x_h) \cdot dx_h \wedge dx_I \wedge dx_J =$$

$$(d_M((\partial f / \partial x_h)) \wedge (g \cdot dx_I \wedge dx_J)) + (-1)^{c(I)} \cdot ((f \cdot dx_I) \wedge (d_M(g \cdot dx_J)))$$

since $dx_h \wedge dx_I = (-1)^{c(I)} \cdot dx_I \wedge dx_h$. Hence (i) of the theorem.

To prove the second part, we can again assume that M is an open set in k^m and that $\alpha = f \cdot dx_I$. Now $d_M(dx_I) = 0$ for any $I \subset [1, m]$. It follows from (i) that we have $d_M(f \cdot dx_I) = \sum_{i \in [1,m]} (\partial f / \partial x_i) \cdot dx_i \wedge dx_I$. It follows that

$$d_M(d_M(f \cdot dx_I)) = \sum_{j \in [1,m]} \sum_{i \in [1,m]} (\partial^2 f / \partial x_j \partial x_i) \cdot dx_j \wedge dx_i \wedge dx_I = 0.$$

Since $dx_j \wedge dx_i = -dx_i \wedge dx_j$ and $\partial^2 f / \partial x_i \partial x_j = \partial^2 f / \partial x_j \partial x_i$.

3.63. Definitions. Set $\Omega^{-1}(M) = \{0\}$. A p-form α on M is *closed* if $d_M(\alpha) = 0$. α is *exact* if there is a β in Ω^{p-1} such that $d_M(\beta) = \alpha$. Thus the set of closed p-foms (denoted $Z^p(M)$ in the sequel) is the kernel of $d_M : \Omega^p(M) \to \Omega^{p+1}(M)$ and is a vector subspace of $\Omega^p(M)$. We denote by $E^p(M)$ the vector subspace $d(\Omega^{p-1}(M))$, i.e., the set of exact p-forms. By (ii) of Theorem 3.61, $E^p(M) \subset Z^p(M)$. The quotient $H^p_{deR}(M) = Z^p(M)/E^p(M)$ is the pth *de Rham cohomology group* of M. If $F : M \to N$ is an analytic map, in view of Corollary 3.59, $F^*(Z^p(N)) \subset Z^p(M)$ and $F^*(E^p(N)) \subset E^p(M)$ so that F^* induces a homomorphism $H^p_{deR}(F)$ of $H^p_{deR}(N)$ in $H^p_{deR}(M)$. Next let $H^p(X, \mathbb{R})$ denote the pth singular cohomology group of a topological space X, and for a continuous map $f : X \to Y$ of X in a topological space Y, let $H^p(f) : H^p(Y) \to H^p(X)$ denotes the induced map of the cohomology groups (see Hatcher [3]).

With this notation, we have the following result.

3.64. Theorem. (Poincaré Lemma) *Let $\varphi = \sum_{I \in \mathcal{I}(p)} f_I \cdot dx_I$ be a p-form on $B_d(0)$ such that $d(\varphi) = 0$. If $p = 0$, φ is a locally constant function. If $p \geq 1$ there is a $(p-1)$-form ψ on $B_d(0)$ such that $d\psi = \varphi$.*

3.65. Proof under an additional assumption. Assume that the following holds:

($*$): the Taylor series $\sum_{\alpha \in \mathbb{N}^m} f_{I\alpha} x^\alpha$ of f_I at 0 converges to f_I uniformly and absolutely on $B_d(0)$.

Let $\Omega^p_*(B_d(0))$ be the space of all p-forms $\omega = \sum_{I \in \mathcal{I}(p)} u_I \cdot dx_I$ where u_I are analytic functions on $B_d(0)$ with their Taylor series at 0 converging uniformly and absolutely to the functions in $B_d(0)$. For $0 \leq p \leq m$ we define the map $H_p : \Omega^p_*(B_d(0)) \to \Omega^{p-1}_*(B_d(0))$ as follows: we set $\Omega^{-1}_*(B_d(0)) = \{0\}$ so that $H_0 = 0$; for $p \geq 1$, we set

$$H_p\left(\sum_{I \in \mathcal{I}(p)} u_I \cdot dx_I \right) = \sum_{I \in \mathcal{I}(p)} \left(\sum_{r=1}^{p} (-1)^r \cdot \tilde{u}_I(r) \cdot dx_{I - \{i_r\}} \right)$$

where $\tilde{u}_I(r) = \sum_{\alpha \in \mathbb{N}^m} (u_{I\alpha}/(|\alpha| + p)) \cdot x^{\alpha + <i_r>}$, with the integer $|\alpha| + p$ treated as an element of k. The Taylor series of $\tilde{u}_I(r)$ converges absolutely and uniformly in $B_d(0)$ as is easily seen.

With this definition of H_p, a straight forward calculation shows that $(d \circ H_p + H_{p+1} \circ d)(\omega) = \omega$ for all ω in $\Omega^p_*(B_d(0))$ if $p \geq 1$ and $(d \circ H_0 + H_1 \circ d)(\omega) = \omega - \omega_0$ if $p = o$ (so that ω is a function). In particular if $d\omega = 0$ and $p \geq 1$, we have $d(H_p(\omega)) = \omega$. When $p = 0$, $\omega = \omega_0$, a constant. This proves the theorem under the additional assumption $(*)$.

3.66. Proof of 3.64 in General for Non-Archimedean k. Let $\Phi = \sum_{I \in \mathcal{I}(p)} f_I \cdot dx_I$ be any p-form. When k is non-archimedean, we can find a finite covering $\{U_i = B_{d_i}(P_i) \subset B_d(0)| \ 1 \leq i \leq q\}$ of $B_d(0)$ such that $U_i \cap U_j = \phi$ for $i \neq j$ and the Taylor series of f_I at P_i converges absolutely and uniformly in $B_{d_i}(P_i)$. By 3.64, for $p \geq 1$ there exist $(p-1)$-forms ψ_i on U_i such that $d\psi_i = \Phi|_{U_i}$. Since the $\{U_i| \ 1 \leq i \leq q\}$ are mutually disjoint, one can define a form ψ on $B_d(0)$ by setting $\psi|_{U_i} = \psi_i$; evidently $d\psi = \Phi$ on all of B_d. Hence the theorem when k is non-archimedean.

3.67. Proof in General for Archimedean k. When k is archimedean, we have an alternative definition for the functions $\tilde{u}_I(r)$ which figure in the definition of the H_p in 3.65: it is easily checked that if $\omega \in \Omega^p_*$,

$$\tilde{u}_I(r)(x) = (-1)^r \cdot x_{i_r} \cdot \int_0^1 u_I(t \cdot x) \cdot t^{p-1} \cdot dt$$

This definition however makes sense for any analytic u_I in $B_d(0)$ whose Taylor series at 0 may not converge in all of $B_d(0)$. Also the function $\tilde{u}_I(r)$ so defined on $B_d(0)$ is analytic in $B_d(0)$.

We will not be needing the Poincaré Lemma in the sequel except in the special case of 1-forms. In the special case of 1-forms, the first two assertions of the Corollary below are restatements of the Poincaré Lemma for 1-forms. The third assertion follows from de Rham's theorem below as $H^1(M, \mathbb{R}) = \{0\}$ for a simply connected manifold M.

3.68. Corollary. *If ω is a 1-form on an analytic manifold M such that $d\omega = 0$, then for a point $p \in M$, there is an open neighbourhood U and an analytic function f on U such that $df = \omega$. If $(U, \Phi : U \to B_d(0))$ is a coordinate system in which $\omega = \sum_{i=0}^m u_i \cdot dx_i$, and the Taylor series at $0 = (f(p))$ of the $\{u_i\}_{1 \leq i \leq m}$, converge uniformly and absolutely in $B_d(0)$, then f can be chosen so that its Taylor series converges in $B_d(0)$. If k is archimedean and M is simply connected, there is an analytic function $f : M \to k$ such that $df = \omega$ (on all of M).*

We state below a celebrated theorem due to G de Rham whose proof is beyond the scope of this book. We will not be using the theorem in the sequel.

3.69. De Rham's Theorem. *Let M be an analytic manifold over $\mathbb{R}$. Then for every $p \in \mathbb{N}$ there is a natural isomorphism $\Phi^p(M)$ of $H^p_{deR}(M)$ on $H^p(M, \mathbb{R})$ such that if $F : M \to N$ is an analytic map, the diagram*

$$
\begin{array}{ccc}
H^p_{deR}(N) & \xrightarrow{H^p_{deR}(F)} & H^p_{deR}(M) \\
\downarrow \Phi^p(N) & & \Phi^p(M) \downarrow \\
H^p(N, \mathbb{R}) & \xrightarrow{H^p(F)} & H^p(M, \mathbb{R})
\end{array}
$$

is commutative i.e., $\Phi^p(M) \circ H^p_{deR}(F) = H^p(F) \circ \Phi^P(N)$.

3.70. Integration on Manifolds. A *volume form* on an analytic manifold M is an m-form ω on M such that $\omega_z \neq 0$ for all $z \in M$. A volume form ω on a manifold M enables one to define a Borel measure on M as follows. Let (U, Φ) be a coordinate chart. We then have the pull back $\Phi^*(\mu_{k^m})$ on U of the measure μ_{k^m} on $\Phi(U)$ (see 1.89). On the other hand the pull back $\Phi^*(dx_{[1,m]})$ of the m-form $dx_{[1,m]}$ (on k^m) under Φ is an m-form which is non-zero at all points of U. It follows that $\omega|_U = \lambda_\Phi \cdot \Phi^*(dx_{[1,m]})$ for an analytic function λ_Φ on U which is non-zero everywhere on U. We define the measure $\mu_{\omega,\Phi}$ on U as the measure $|\lambda_\Phi|_k \cdot \Phi^*(\mu_{k^m})$ when k is non-archimedean or is $\mathbb{R}$; when $k = \mathbb{C}$, we define $\mu_{\omega,\Phi}$ on U as $|\lambda_\Phi|^2_k \cdot \Phi^*(\mu_{k^m})$ where μ_{k^m} is a fixed Haar measure on k^m. When $m = 1$, one takes μ_k as the Haar measure for which $\mu(\{t \in k| \ |t| \leq 1\}) = 1$ (resp. 2, resp. π) if k is non-archimedean (resp. $\simeq \mathbb{R}$, resp. $\simeq \mathbb{C}$). For $n \geq 2$, μ_{k^n} is the n-fold product of μ_k with itself. We then have the following.

3.71. Lemma. *Let $(U, \Phi), (U', \Phi')$ be two coordinate charts on M. Then for any Borel set $E \subset U \cap U'$, $\mu_{\omega,\Phi}(E) = \mu_{\omega,\Phi'}(E)$.*

3.72. The lemma is an immediate consequence of the change of variable formula for integration on open sets k^m. The formula is proved in many undergraduate texts in the case $k = \mathbb{R}$. The proof given in Narasimhan [8, Chapter 5, Theorem 1] (in the case $k = \mathbb{R}$) carries over to the case of non-archimedean k with minor modifications. It enables one to define (by patching the $\mu_{\omega,\Phi}$ as the (U, Φ) as they vary over an atlas) a Borel measure μ_ω on M. When k is non-archimedean, we have noted that any analytic manifold M has an atlas $\{(U_i, \varphi_i)| \ i \in I\}$ such that $U_i \cap U_j = \phi$ if $i \neq j$. It follows that any analytic manifold over a non-archimedean k admits a volume form.

3.73. Archimedean k. The situation when k is archimedean is more complicated. Not all manifolds over $\mathbb{R}$ admit a volume form. The manifold M over $\mathbb{R}$ is said to be *orientable* if it admits a volume form. Let $\mathfrak{V}$ be the set of all volume forms on M. Define the relation $\sim$ as follows: for $\omega, \omega' \in \mathfrak{V}$, $\omega \sim \omega'$ if

$\omega' = f \cdot \omega$ with $f(z) > 0$ for all $z \in M$. Evidently $\sim$ is an equivalence relation. An orientation on M is an equivalence class in $\mathfrak{V}/\sim$. If ω is a volume form so is $-\omega$, but they are not (evidently) equivalent. The orientation determined by $-\omega$ is the *opposite* of the orientation determined by ω. An example of a manifold which is not orientable is $\mathbb{P}^n(\mathbb{R})$ (see 3.10 (6)) for n even. We have the following.

3.74. Theorem. *The following conditions on an analytic manifold over $\mathbb{R}$ are equivalent:*
(i) *M is orientable.*
(ii) *M admits an atlas $\{(U_i, \Phi_i)|\ i \in I\}$ such that for every $i, j \in I$, $\det(d\Phi_i \circ \Phi_j^{-1})_{\Phi(z)} > 0$ for all $z \in (U_i \cap U_j)$.*

3.75. The implication (i)$\Rightarrow$ (ii) is easy to prove and is left as an exercise to the reader. (ii)$\Rightarrow$(i) however depends on some deep results on analytic manifolds over $\mathbb{R}$ and so we do not give a proof. We will now use condition (ii) of the theorem to show the following.

3.76. Proposition. *A complex analytic manifold M treated as a real analytic manifold is orientable.*

3.77. Proof. Let $\{(U_i, \Phi_i : U_i \to \mathbb{C}^m)|\ i \in I\}$ be a complex analytic atlas on M. Let $T : \mathbb{C}^m \to \mathbb{R}^{2 \cdot m}$ be the $\mathbb{R}$-vector space isomorphism defined as follows: for $z = \{z_r = x_r + i \cdot y_r|\ 1 \le r \le m\}$ $(x_r, y_r \in \mathbb{R})$, $T(z) = \{u_s\}_{1 \le s \le 2 \cdot m}$ where $u_{(2q-1)} = x_q$ and $u_{2 \cdot q} = y_q$ for $1 \le q \le m$. Then $\{(U_i, T \circ \Phi_i)|\ i \in I\}$ is an atlas for M as a real analytic manifold. For this atlas we have $\det(d(T \circ \Phi_i) \circ d(T \circ \Phi_j)^{-1}_{(T \circ \Phi_j)(z)}) = |\det\{(\Phi_i \circ \Phi_j^{-1})\}_{\Phi(z)}|^2$ showing that (ii) holds for M as a real analytic manifold.

4. Lie Groups

Lie groups are named after the Norwegian mathematician Sophus Lie (1842–1899), who laid the foundations of the theory of continuous transformation groups. They were christened Lie groups by Arthur Tress, a student of Lie. Lie proved the basic result which sets up a correspondence between Lie Groups and Lie Algebras which we call the Fundamental Theorem of Lie Theory. This reduces most problems about Lie groups to (essentially algebraic) questions about Lie algebras. Wilhelm Killing and Elie Cartan made giant strides in developing the theory. The modern treatment of Lie groups is due to Claude Chevalley whose 1950 book on Lie groups is a classic. All text-books on Lie groups (including this one) essentially follow his treatment.

4.1. A *Lie group* over (a locally compact field) k is an analytic manifold G over k, together with an analytic map $\mu : G \times G \to G$, which is a group structure on the underlying set of G. We will see that this implies the analyticity of $\mathcal{I}v$, the map of G in G that takes each $g \in G$ to its inverse g^{-1} . We denote the dimension of G as an anlytic manifold by m. For $x, y \in G$, $\mu(x, y)$ is also denoted $x \cdot y$. Evidently, G is a locally compact topological group. If $k' \subset k$ is a locally compact subfield of k, G has a natural structure $R_{k/k'}G$, of an analytic manifold over k' (see 3.6) and one verifies easily that the map μ is analytic for the analytic structure over k' as well. In the sequel when we speak of a Lie group G over k as a Lie group over k' we mean the Lie group $R_{k/k'}G$ (over k').

4.2. In the rest of this book G will denote a Lie group. Let $(U, \Theta : U \to B_d(0))$ (cf. 1.92 for notation) be a chart on G at 1 with $\Theta(1) = 0$. Let $0 < d' \leq d$ satisfy the following conditions:
(i) If $V = \Theta^{-1}(B_{d'}(0))$, then $\mu(V \times V) \subset U$.
(ii) The Taylor series at $(0,0)$ of $F = \Theta \circ \mu \circ (\Theta^{-1} \times \Theta^{-1})$ on $B_{d'}(0) \times B_{d'}(0)$ converges to F absolutely and uniformly in $B_{d'}(0) \times B_{d'}(0)$.

The fact that $\Theta^{-1}(0) = 1$ leads us immediately to the conclusion that $F((0,0)) = 0$ and $F(0, x) = F(x, 0) = x$ for all $x \in B_{d'}(0)$. Thus the Taylor expansion of F at $(0,0)$ is of the form

$$F(x, y) = x + y + \sum_{(\alpha,\beta) \in (\mathbb{N}^m \times \mathbb{N}^m), |\alpha| + |\beta| \geq 2} F_{(\alpha,\beta)} \cdot x^\alpha \cdot y^\beta \qquad (*)$$

with $F_{(\alpha,\beta)} \in k^m$ for $(x,y) \in B_{d'}(0) \times B_{d'}(0)$. $(*)$ implies the following.

4.3. Proposition. *The differential* $d\mu_{(1,1)} : T_1G \times T_1G \to T_1G$ *of* μ *at* $(1,1)$ *takes* $(v,w) \in T_1G \times T_1G = T_{(1,1)}(G \times G)$ *to* $v + w$ *in* T_1G.

4.4. Corollary. *The map* $\mathcal{I}v : G \to G$ *is analytic.*

4.5. Proof. For $x \in B_{d'}(0)$ the inverse of $g = \Theta^{-1}(x)$ (if it is in V) is $\Theta^{-1}(y)$ where $y \in B_d(0)$ satisfies the equation $F(x,y) = 0$. Now $F(0,0) = 0$ and the differential of $F(0,\cdot)$ at 0 is surjective. By the Implicit Function Theorem there is an open set $\Omega \ni 0$ in $B_d(0)$ and an analytic $u : \Omega \to V$ such that $F(x,y) = 0$ for $x \in \Omega$ if and only if $y = u(x)$. It follows that on the open set $W = \Theta^{-1}(\Omega)$, $\mathcal{I}v = \Theta^{-1} \circ u \circ \Theta$ is analytic. As $(g \cdot h)^{-1} = h^{-1} \cdot g^{-1}$ for $h \in W$ and translations by g are analytic, we see that $\mathcal{I}v$ is analytic on G.

4.6. Remarks. (1) Replacing $d' > 0$ (in 4.2) by a smaller constant if necessary, we see that the Taylor series of the function $u = \Theta \circ \mathcal{I}v \circ \Theta^{-1}$ is absolutely and uniformly convergent in $\Omega'(= B_{d'}(0))$.

(2) Given any $d > 0$, there is a $\lambda \in k$ such that for $x \in k^m$, $B_d(x) = B_{|\lambda|_k}(x)$. In the sequel *all open discs* $B_d(x)$ *considered will be such that* $d = |\lambda|_k$ *for some* $\lambda \in k$; hence $\lambda^{-1} \cdot B_d(x) = B_1(x)$.

(3) Assume that k is non-archimedean. The Taylor series $(*)$ being convergent, $\exists\, M, r > 0$ such that

$$||F_{(\alpha,\beta)}||_k \le M \cdot r^{-(|\alpha+\beta|)}$$

for all $(\alpha,\beta) \in \mathbb{N}^m \times \mathbb{N}^m$. Further there is a $0 < d_1 < \min\{r,d\}$ such that the Taylor series of F at $(0,0)$ converges to F absolutely and uniformly in $B_{d_1}(0) \times B_{d_1}(0)$. Now, from $(*)$, one has for $x,y \in B_{d_1}(0)$,

$$||F(x,y)||_k \le \max(d_1, \{M \cdot (d_1/r)^{|\alpha+b|}| \; \alpha,\beta \in \mathbb{N}^m, |\alpha + \beta| \ge 2\}) \le d_1,$$

if $\max(M \cdot d_1/r, d_1/r) \le 1$. Thus if d' in 4.2 is sufficiently small, $F(B_{d'}(0) \times B_{d'}(0)) \subset B_{d'}(0)$. Thus we have proved the following.

4.7. Theorem. *If* k *is non-archimedean,* G *admits a fundamental system* $\{U_n| \; n \in \mathbb{N}\}$ *of neighbourhoods of* 1, *which are compact open subgroups of* G. *In fact, if* $(U, \Theta : U \to B_{d_1}(0))$ *is any chart with* $\Theta(1) = 0$, *then there is a* $0 < d_2 \le d_1$ *such that* $\Theta^{-1}(B_d(0))$ *is an open compact subgroup of* G *for all* $0 < d \le d_2$.

On the other hand, for archimedean k, we have the following.

4.8. Theorem. *Let* $k = \mathbb{R}$ *or* $\mathbb{C}$. *Then the identity connected component* G^o *of a Lie group* G *over* k *is an open and closed subgroup of* G *which is generated by every neighbourhood of* 1 *contained in* G^o.

4.9. Proof. That G^o is open and closed in G follows from the fact that G is locally (arc-wise) connected. Let V be a neighbourhood of 1 in G contained in G^o and G', the subgroup generated by V. Then for any $g \in G'$, the neighbourhood $g \cdot V$ of g in G, is contained in G'. Thus G' is open and hence closed in G. As V is contained in G^o, $G' = G^o$.

4.10. Examples. Here are some Lie groups that make their appearance naturally in several contexts in mathematics. That the underlying space has a natural structure of an analytic manifold in which the group operation is analytic, is evident in cases (**1**)–(**6**) below. In the other cases we have briefly indicated how the underlying space may be given an analytic structure leaving it to the reader to supply details.

(**1**) The trivial group is evidently a Lie group. Any countable group with the discrete topology is evidently a Lie group (of dimension 0).

(**2**) The simplest (non-trivial) example of a Lie group is the k under addition. More generally, any finite dimensional vector space V over k is a Lie group under addition: the analytic structure on V is obtained by identifying V with k^n through a basis; it is independent of the basis as linear maps from k^p to k^q are analytic.

(**3**) $k^\times$ with multiplication is a Lie group – $k^\times$ is an open set in k. For an integer $n > 0$, the general linear group $GL(n, k)$ of non-singular $(n \times n)$-matrices (under matrix multiplication) is a Lie group – $GL(n, k) = \{g \in M(n, k) \mid \det(g) \neq 0\}$ is an open subset of $M(n, k)$ and matrix multiplication is a quadratic polynomial map of $M(n, k) \times M(n, k)$ into $M(n, k)$. Note that $GL(1, k)$ is the same as $k^\times$.

(**4**) If G, H are Lie groups, then $G \times H$ with the direct product structure for the group as well as the analytic structure.

(**5**) The subgroup of diagonal matrices in $GL(n, k)$ is a Lie group – it is obviously the same as the n-fold product $(k^\times)^n$ of $k^\times$ with itself.

(**6**) The sets $S = \{g \in GL(n, k) \mid g_{i,j} = 0 \text{ } for \text{ } i > j\}$ and $U = \{g \in S \mid g_{ii} = 1\}$ are evidently subgroups of $GL(n, k)$. The map $F : U \to k^{n(n-1)/2}$, defined by setting $F(g) = \{g_{i,j}\}_{1 \leq i < j \leq n}$, identifies U with $k^{n(n-1)/2}$ while the bijection $\tilde{F} : D \times U \to S$, defined by setting $\tilde{F}(x, u) = x \cdot u$ for $(x, u) \in D \times U$, gives an analytic manifold structure on S; and they are Lie groups under these analytic structures. The inclusions of S and U in $GL(n, k)$ make them (closed) analytic sub-manifolds of $GL(n, k)$ (see 3.25).

(**7**) $SL(n, k) = \{g \in GL(n, k) \mid \det(g) = 1\}$ has a natural sructure of a Lie group. Hint: apply the Implicit Function Theorem to the k-valued analytic function $A \rightsquigarrow \det(A)$ on $M(n, k)$.

(8) Let $O(n) = \{g \in GL(n,k)|\ {}^t g \cdot g = 1\}$. Then evidently $O(n)$ is a subgroup of $GL(n,k)$. It is compact (exercise). One has a natural structure of an analytic manifold of dimension $n(n-1)/2$ on $O(n)$: to see this let $\mathfrak{st}(n)$ denote the vector space of skew symmetric matrices in $M(n,k)$ and $\mathcal{B}$ be the open subset $\{g \in \mathfrak{st}(n)|\ (g \pm 1)\ are\ invertible\}$ of $\mathfrak{st}(n)$; then $\mathcal{B}$ is non-empty: it suffices to check this in the cases $n = 1$ and 2 (exercise). Then the map $\mathcal{C}'$ of $\mathcal{B}$ taking $T \in \mathcal{B}$ to $(1-T)(1+T)^{-1}$, is an analytic map of $\mathcal{B}$ into $GL(n,k)$ mapping $\mathcal{B}$ homeomorphically onto the open set $\mathcal{B}^* = \{A \in O(n)|\ (A \pm 1)\ are\ invertible\}$ in $O(n)$: the inverse of this map which takes $A \in \mathcal{B}^*$ to $(1-A)(1+A)^{-1}$ which we denote $\mathcal{C}$ is the *Cayley Transform*. The collection $\{(L_g(\mathcal{B}^*), \mathcal{C} \circ L_g^{-1})|g \in G\}$ is an analytic atlas on $O(n)$ which makes $O(n)$ into a Lie group. We leave it to the reader to check these assertions.

(9) Let G be a Lie group (over k) and Γ be a subgroup of G such that the topology induced on Γ by that on G is discrete. Then Γ is closed in G (exercise). Let $V' \ni e$ be an open set in G with $V' \cap \Gamma = \{e\}$. Let $V \ni e$ be an open set in G such that $V^{-1} \cdot V = \{g^{-1} \cdot h|\ g, h \in V\}$ is contained in V'; then $V \cdot \gamma \cap V \neq \phi$ for $\gamma \in \Gamma$ if and only if $\gamma = e$. Let (U, Θ) be a coordinate chart on G with $e \in U \subset V$. Set $M = G/\Gamma$ with the quotient topology and $\pi : G \to M$ the natural map. Then $\pi|_U$ is a homeomorphism of U onto the open set $\pi(U)$. The collection $\{(g \cdot \pi(U), \Theta \circ (\pi|_U)^{-1} \circ L_g^{-1})|\ g \in G\}$ is an analytic atlas on M. With this definition, the natural map $G \to M = G/\Gamma$ is analytic as also the map $\bar{\mu} : G \times M \to M$ defined by setting for $(x, y \cdot \Gamma) \in G \times M$, $\bar{\mu}(x, y \cdot \Gamma) = x \cdot y \cdot \Gamma$. When Γ is a normal subgroup, $\bar{\mu}$ factors through $G/\Gamma \times G/\Gamma$ to give a group structure on G/Γ making it a Lie group.

(10) Let $k = \mathbb{R}$ (or $\mathbb{C}$) and G a connected Lie group. G being locally contractible admits a universal covering $u_G : \tilde{G} \to G$. Moreover $\tilde{G}$ has a natural structure of a topological group such that $u_G : \tilde{G} \to G$ is a continuous group homomorphism. The kernel Γ of u_G is a discrete normal subgroup and is abelian (see 1.81). One defines an atlas on $\tilde{G}$ as follows. Let $\mathcal{U} = \{(U_i, \Theta_i : B_d(0))|\ i \in I\}$ be an atlas for G with $\Theta_i(U_i)$ the unit disc in k^m. Then for each $i \in I$, $p^{-1}(U_i)$ is a disjoint union $\coprod_{\gamma \in \Gamma} U_{i\gamma}$ of open subsets of $\tilde{G}$, and $p|_{U_{i\gamma}} : U_{i\gamma} \to U_i$ is a homeomorphism onto U_i. Then $\{(U_{i\gamma}, (\Theta_i \circ p))|\ (i, \gamma) \in I \times \Gamma\}$ is an atlas on $\tilde{G}$. The covering map u_G as well as the multiplication $\tilde{\mu}$ on $\tilde{G}$ deduced from that on G, are easily seen to be analytic.

(11) Suppose that H and G are Lie groups and we are given an analytic group action of H on G. This means that we are given an analytic map $\Phi : H \times G \to G$ such that
(i) $\Phi(h \cdot h', g) = \Phi(h, \Phi(h', g))$ for $h, h' \in H$ and $g \in G$ and

(ii) $\Phi(h, g \cdot g') = \Phi(h, g) \cdot \Phi(h, g')$ for $h \in H$ and $g, g' \in G$.

Condition (ii) says that for fixed $h \in H$, if $\varphi(h) : G \to G$ is defined by setting for $g \in G$, $\varphi(h)(g) = \Phi(h, g)$, $\varphi(h)$ is an analytic group automorphism of G. Consider the semi-direct product $H \propto_\varphi G$ as discrete groups (cf. (1.13)). The underlying set of $H \propto_\varphi G$ is $H \times G$ which has the product analytic structure. With this analytic structure, $H \propto_\varphi G$ is a Lie group. Note that the direct product $H \times G$ is the special case when $\varphi(h)(g) = g$ for all $g \in G$ and $h \in H$.

4.11. Left and Right Translation Invariant Vector Fields. For $g \in G$, the left (resp. right) translation $L_g : G \to G$ (resp. $R_g : G \to G$) is an analytic diffeomorphism. L_g (resp. R_g) defines an isomorphism of the vector space $\mathcal{T}(G)$ of analytic vector fields on itself which too we denote by L_g (resp. R_g). A vector field X on G is *left* (resp. *right*) *translation invariant* if $L_g(X) = X$ (resp. $R_g(X) = X$). From the analyticity of μ, it is easily deduced that for $v \in T_e(G)$, the assignment $g \rightsquigarrow dL_g(v)$ denoted v_L (resp. $g \rightsquigarrow dR_g(v)$ denoted v_R) is a left (resp. right) translation invariant analytic vector field on G. Moreover, as is easy to see, every left (resp. right) translation invariant vector field X on G equals $(X_e)_L$ (resp. $(X_e)_R$). Thus the vector space $\mathcal{L}(G)$ (resp. $\mathcal{R}(G)$) of all left (resp. right) translation invariant vector fields on G is isomorphic as a vector space to T_eG. Since $L_g([X, Y]) = [L_g(X), L_g(Y)]$ (resp. $R_g([X, Y]) = [R_g(X), L_g(Y)]$), $\mathcal{L}(G)$ (resp. $\mathcal{R}(G)$) is a Lie sub-algebra of $\mathcal{T}(G)$. The Lie algebra automorphism $\mathcal{I}v_*$ of $\mathcal{T}(G)$ induced by $\mathcal{I}v$ gives an isomorphism of the Lie algebra $\mathcal{L}(G)$ on $\mathcal{R}(G)$. Thus the tangent space at e to G has a natural structure of a Lie algebra.

4.12. The Lie Algebra of G. The Lie algebra $\mathcal{L}(G)$ (also denoted Lie(G) or $\mathfrak{g}$) defined above is the *Lie algebra of G*. It is evidently of dimension $m = \dim_k G$ over k. In general, in this book, Lie groups are denoted by capital Roman letters and their Lie algebras by the corresponding lower case Gothic letters. We saw that the underlying vector space of $\mathfrak{g}$ is naturally isomorphic to the tangent space $T_e(G)$ to G at 1 and we identify $T_e(G)$ with $\mathfrak{g}$ through this isomorphism and treat the tangent space at 1 itself as the Lie algebra $\mathfrak{g}$. As there is a canonical isomorphism of $\mathcal{R}(G)$ on $\mathcal{L}(G)$, so the Lie algebra of G could have also been defined as $\mathcal{R}(G)$. If k' is a locally compact subfield of k, as was noted earlier, G may be regarded as a Lie group over k'. From the definitions it is clear that $\mathfrak{g}$ regarded as a Lie algebra over k' is the Lie algebra of G regarded as a Lie group over k'. If $f : G \to H$ is an analytic map with $f(1) = 1$, it induces a k-linear map $df_1 : T_1(G) \to T_1(H)$. When f is a group homomorphism, as we will see, df_1 is a Lie algebra homomorphism of $\mathfrak{g}$ in $\mathfrak{h}$ $(= \text{Lie}(H))$. But before that we will determine the Lie algebras of two special Lie groups.

4.13. Theorem. *The Lie algebra of a vector space V over k considered as a Lie group over k is V with the trivial bracket operation.*

4.14. Proof. If we identify V with k^m through a basis, the vector fields $\{\partial/\partial x_i | 1 \leq i \leq m\}$ are left (and right) translation invariant vector fields on V; evidently they constitute a basis for the Lie algebra $\mathfrak{v}$ of V and thus $\mathfrak{v}$ gets identified with V itself. Further $[\partial/\partial x_i, \partial/\partial x_j] = 0$ for all $1 \leq i,j \leq m$, so the bracket operation on $\mathfrak{v}$ is trivial.

4.15. Theorem. *The Lie algebra of the Lie group $GL(n,k)$ is naturally isomorphic to $\mathfrak{gl}(n,k)$ (see 1.45 for notation).*

4.16. Proof. For $1 \leq i,j \leq n$ let $E_{i,j}$ denote the matrix whose $(ij)^{\text{th}}$ entry is 1 and all other entries are 0. Then $\{E_{i,j} | 1 \leq i,j \leq n\}$ is a basis of $\mathfrak{gl}(n,k) = M(n,k)$. Let $\{x_{i,j} | 1 \leq i,j \leq n\}$ be the coordinate system on $\mathfrak{gl}(n,k)$ as also on the open subset $GL(n,k)$ given by this basis. Then the left translation invariant vector field $X_{i,j}$ on $GL(n,k)$ which at 1_n equals $\partial/\partial x_{i,j}$ is easily seen to be $\sum_{q=1}^{n} x_{qi} \cdot (\partial/\partial x_{qj})$. Further, it is easily checked that $[X_{i,j}, X_{kl}] = \delta_{jk} \cdot X_{il} - \delta_{li} \cdot X_{lj}$. As $\{X_{i,j} | 1 \leq i,j \leq n\}$ is a basis of the Lie algebra of $GL(n,k)$ we see that this Lie algebra is isomorphic to $\mathfrak{gl}(n,k)$.

4.17. The 1-Parameter Group of a Left Invariant Vector Field. In 3.39, we saw that given an analytic vector field X on G and a coordinate chart $(U, \Theta : U \to B_d(0))$ at 1 (so that $1 \in U$ and $f(1) = 0$) with $\bar{U}$ compact, there is a $\delta > 0$ and an analytic map $\Phi : I_\delta(0) \times U \to G$ with the following properties:

(**i**) $\Phi(0,x) = x$ for all $x \in U$.

(**ii**) $\exists\, b > 0, b \leq d$ such that if $U' = \Theta^{-1}(B_b(0))$, $\Phi(I_\delta(0) \times U') \subset U$.

(**iii**) For each $t \in I_\delta(0)$, the map $x \rightsquigarrow \Phi(t,x)$ is an analytic diffeomorphism of U' onto its (open) image $\Phi(t, U')$.

(**iv**) If $t, t', (t+t') \in I_\delta(0)$ and x as well as $\Phi(t,x)$ are in U, then $\Phi(t', \Phi(t,x)) = \Phi(t + t', x)$.

(**v**) For an analytic function f defined on an open subset W of U, one has for all $x \in W$, $d/dt(f(\Phi(t,x)))|_{t=0} = (Xf)(x)$.

(**vi**) The Taylor series of the function $\Theta \circ \Phi \circ (1_{I_\delta(0)} \times \Theta^{-1})$ (f as in **v**) at any point of $I_\delta(0) \times B_b(0)$ converges to the function uniformly and absolutely on all of $I_\delta(0) \times B_b(0)$.

4.18. Suppose now that X is a left translation invariant vector field on G and $(\Phi U, \delta, b)$ are chosen satisfying (i)–(v) above. Let V be a neighbourhood of the identity such that $V \cdot V^{-1} = \{v \cdot w^{-1} | v, w \in V\} \subset U$.

When k is non-archimedean, we assume, as we may, that V is a compact open subgroup and δ (replaced by a smaller constant if need be) is such that $\Phi(t,x) \in V$ for all $t \in I_\delta$ and $x \in V$ – note that the set $\{t \in I_\delta |\ \Phi(t,V) \subset V\}$ is open and contains 0. It follows that there is a countable subset $S = \{g_n |\ n \in \mathbb{N}\}$ of G such that G is the disjoint union of $\{g_n \cdot V |\ n \in \mathbb{N}\}$ (cosets modulo V). Define $\Phi_n : I_\delta(0) \times g_n \cdot V \to g_n \cdot V$ by setting for $t \in I_\delta(0)$ and $x \in V$, $\Phi_n(t, g_n \cdot x) = g_n \cdot \Phi(t,x)$ and $\Phi_X : I_\delta(0) \times G \to G$ by $\Phi_X|_{I_\delta(0) \times g_n \cdot V} = \Phi_n$. Evidently Φ is analytic and we have for $g \in G$ and

$t, t' \in I_\delta(0)$, $\Phi_X(t+t', g) = \Phi_X(t, \Phi_X(t', g))$. Thus we see that the 1-parameter group of the vector field X is defined on all of G and for all $t \in I_{\delta(0)}$ even if G is not compact.

A statement analogous to the last one holds also in the case of archimedean k. To see this we need the following.

4.19. Proposition. *Let k be archimedean. Let U, δ, b be as in 4.17. For $g \in G$, let $\Phi_g : I_\delta(0) \times g \cdot V \to G$ be the analytic map defined by setting for $t \in I_\delta(0)$ and $x \in V$, $\Phi_g(t, g \cdot x) = g \cdot \Phi(t, x)$. Then for $g, g' \in G$ and $h \in (g \cdot V) \cap (g' \cdot V)$, $\Phi_g|_{I_\delta(0) \times (g \cdot V) \cap g' \cdot v} = \Phi_{g'}|_{I_\delta(0) \times (g \cdot V) \cap g' \cdot V}$.*

4.20. Proof. For fixed h, both $\Phi_g|_{I_\delta(0) \times \{h\}}$ and $\Phi_{g'}|_{I_\delta(0) \times \{h\}}$ are analytic solutions of a boundary value problem in Ordinary Differential Equations with the same initial conditions and hence their germs at $t = 0$ coincide. Since $I_\delta(0)$ is connected, by the principle of analytic continuation they are equal on all of $I_\delta(0) \times h$. Since $h \in (g \cdot V) \cap (g' \cdot V)$ is arbitrary, this proves the proposition.

The proposition clearly implies the Assertions (i)–(iii) in the following

4.21. Theorem. *Let G a Lie group over k. Let X be a left translation invariant vector field on G. Then there is a $\delta > 0$ and an analytic map $\Phi_X : I_\delta(0) \times G \to G$ satisfying the following conditions:*
(i) $\Phi_X(0, g) = g$, $\forall g \in G$.
(ii) $\Phi_X(t, \Phi_X(t', g)) = \Phi_X(t + t', g)$, $\forall g \in G$ and $t, t', t + t' \in I_\delta(0)$.
(iii) *For an analytic function u defined in the neighbourhood of any point $g \in G$, $d/dt(u \circ \Phi_X(t, g))|_{t=0} = (X_g \cdot u)$.*
When k is non-archimedean $I_\delta(0)$ is a group and the condition on t, t' in (ii) is redundant.
When k is archimedean Φ_X can be defined on $k \times G$ with (i)–(iii) holding for all $t, t' \in k$ and $g \in G$.

4.22. Proof. The only unproved assertion is the last one (for archimedean k). Set for $g \in G$ and $t \in I_\delta(0)$, $\Phi_{Xt}(g) = \Phi_X(t, g)$ so that Φ_{Xt} is an analytic diffeomorphism of G on itself. We take up the case $k = \mathbb{R}$ first. Now any element of $t \in \mathbb{R}$ is uniquely of the form $n \cdot \delta/2 + \xi$ where $n \in \mathbb{Z}$ and $0 \le \xi < \delta/2$. We define $\Phi_{Xt}(g)$ as $(\Phi_{X\delta/2})^n \circ \Phi_{X\xi}$. When $k = \mathbb{C}$, any $t \in \mathbb{C}$ is uniquely of the form $(m + i \cdot n) \cdot \delta/2 + (\xi + i \cdot \eta)$ with $m, n \in \mathbb{Z}$, $\xi, \eta \in \mathbb{R}$ with $0 \le \xi, \eta < \delta/2$. One then defines $\Phi_{Xt}(g)$ as $(\Phi_{X\delta/2})^m \circ (\Phi_{Xi \cdot \delta/2})^n \circ \Phi_{X\xi+i\cdot\eta}(g)$. It is easily checked that (ii) and (iii) hold for any $g \in G$ and $t, t' \in k$.

4.23. Next observe that we may regard $\mathfrak{g}$ as a family of vector fields on G (parametrized by $\mathfrak{g}$): each $X \in \mathfrak{g}$ defines a vector field on G (also denoted X). Applying to this family of vector fields Theorem 3.37 combined with Proposition 4.19, we obtain the following.

4.24. Theorem.   *Given any relatively compact open subset Ω in $\mathfrak{g}$ there is a $\delta > 0$ and an analytic map $\Phi : \Omega \times G \times I_\delta \to G$ satisfying the following conditions:*
(i) *$\Phi(X, g, 0) = g$ for all $X \in \Omega$ and $g \in G$.*
(ii) *$\Phi(X, (\Phi(X, g, s), t)) = \Phi(X, g, t + s)$ for all $X \in \Omega$, $t, s \in I_\delta(0)$ with $t + s \in I_\delta(0)$.*
(iii) *If $X \in \Omega$ and $\lambda \in k$ is such that $\lambda \cdot X \in \Omega$, $\Phi(\lambda \cdot X, g, t) = \Phi(X, g, \lambda \cdot t)$ for $t \in I_\delta(0)$ with $\lambda \cdot t \in I_\delta(0)$.*
(iv) *For any analytic k-valued function u defined in a neighbourhood of $g \in G$ and $X \in \mathfrak{g}$,*

$$(X \cdot u)(\Phi(X, g, t)) = d(u(\Phi(X, g, t)))/dt.$$

When k is archimedean, δ may be taken as ∞, i.e., Φ is defined on all of $\mathfrak{g} \times G \times k$.

4.25. Remarks.   We record here some observations about the analytic function Φ. We denote $\Phi(X, g, t)$ also by $\varphi_{X,t}(g)$.

(**1**) (ii) Can be formulated as $\varphi_{X,t+t'} = \varphi_{X,t} \circ \varphi_{X,t'}$.

(**2**) When k is non-archimedean, $I(\delta)$ is a subgroup of k, so that in (ii), if $t, s \in I_\delta(0)$, automatically $t + s \in I_\delta(0)$.

(**3**) Let k' be a locally compact subfield of k. If $\{\varphi_{X,t} : G \to G| \, t \in I_\delta\}$ is the 1-parameter group of the vector field $X \in \mathfrak{g}$, then $\{\varphi_{X,t} : G \to G| \, t \in I_\delta \cap k'\}$ is the 1-parameter group of X considered as a vector field on $R_{k/k'}G$ i.e., on G as a Lie group over k' (see 3.40).

(**4**) The uniqueness assertion in the existence theorem for ordinary differential equations tells us that the germ of Φ at (X, g, t) is unique. When k is archimedean and G is connected) the principle of analytic continuation ensures that Φ satisfying (i)–(iv) in the theorem is unique on all of $\mathfrak{g} \times k \times G$.

(**5**) When k is non-archimedean, the global uniqueness in (4) fails as is shown by the following example. Let $\varpi : \mathfrak{o}_k \to F$ be the natural map of the integers in k onto the residue field F. Let $G = \mathfrak{o}_k \times F$; the Lie algebra of G is then k (with the trivial bracket operation). Define $\Phi : \mathfrak{o}_k \times \mathfrak{o}_k \times G \to G$ (resp. $\Phi' : \mathfrak{o}_k \times \mathfrak{o}_k \times G \to G$) as follows: for $X \in \mathfrak{o}_k (\subset$Lie algebra of G), $t \in \mathfrak{o}_k$ and $(x, \xi) \in \mathfrak{o}_k \times F(= G)$, even if G is not

$$\Phi(X, t, (x, \xi)) = (x + t \cdot X, \xi)$$
$$(\text{resp. } (\Phi'(X, t, (x, \xi)) = (x + t \cdot X, \xi) + \pi(t \cdot X))).$$

Then Φ and Φ' coincide on $\mathfrak{o}_k \times \mathfrak{p}_k \times G$ and so have the same germ at all points in $\mathfrak{o}_k \times \mathfrak{p}_k \times G$ and both satisfy the conditions in the theorem for $t, s \in \mathfrak{o}_k(= I_1(0))$, $X \in \mathfrak{o}_k$ and $g \in G$; but $\Phi \neq \Phi'$.

4.26. Lemma. *Let Ω be a relatively compact open subset of $\mathfrak{g}$. Let $X \in \Omega$ be a left translation invariant vector field on G. Let δ and $\{\varphi_{Xt}\}_{t \in I_\delta}$ be as in 4.24 above. Set $J = k$ or $I_\delta(0)$ according to whether k is archimedean or not. Then there is an analytic group homomorphism $E_X : I_\delta(0) \to G$ such that $\varphi_{Xt}(g) = g \cdot E_X(t)$ for all $t \in J$. Further the map $(t, X) \rightsquigarrow E_X(t)$ is an analytic map of $\Omega \times I_\delta(0)$ in G. Moreover for $\lambda \in k$, one has $E_{\lambda \cdot X}(t) = E_X(\lambda \cdot t)$ if both $(X, \lambda \cdot t)$ and $(\lambda \cdot X, t)$ are in $\Omega \times J$.*

4.27. Proof. Since X is left translation invariant we see that the analytic diffeomorphism $\varphi_X, t : G \to G$ necessarily commutes with all left translations; and this means that it is necessarily right multiplication by the element $E_X(t) = \varphi_X, t(1)$. It follows that $(X, t) \rightsquigarrow E_X(t)$ is analytic in both t and X. That E_X is a homomorphism of J in G is immediate from the fact that $\varphi_{X(t+t')} = \varphi_{Xt} \circ \varphi_{Xt'}$.

4.28. The Exponential Map. The function $E_X(t)$ is an analytic function defined on an open set of $\mathfrak{g} \times k$ of the form $B_{d'}(0) \times I_{\delta'}(0)$. Let $0 \neq u \in k$ be such that $|u|_k = \rho < \delta'$ for some $u \in k$. Set $d = \rho \cdot d'$ and define $\{\exp_G : B_d(0) \to G$, by setting for $X \in B_d(0)$,

$$\exp_G(X) = E_{u^{-1} \cdot X}(u)$$

One then checks that if $Y = t \cdot X$ with $t \in k < |t| \leq 1$, $\exp_G(Y) = E_{\rho^{-1} \cdot X}(\rho \cdot t)$ and that for $X \in B_d(0)$ and $t, s \in k$ with $|t|, |s|, |t+s| < 1$,

$$\exp_G((t + s) \cdot X) = \exp_G(t \cdot X) \cdot \exp_G(s \cdot X).$$

Suppose that k is archimedean. Then $\delta = \infty$ and $I_\delta(0) = k$. We then define $\exp_G : \mathfrak{g} \to G$ the *exponential map* (of G) by setting for $X \in \mathfrak{g}$, $\exp_G(X) = E_X(1)$. We have then

$$\exp_G(t \cdot X) = E_X(t) \ for \ all \ t \in k. \qquad (*)$$

The following proposition is an easy consequence of the definition of the exponential map and the inverse function theorem.

4.29. Proposition. *Under the identification of $T_1(G)$ with $\mathfrak{g}$, and of the tangent space at 0 to $\mathfrak{g}$ also with $\mathfrak{g}$, the differential of the exponential map (defined on $B_d(0)$ for some $d > 0$) at $0 \in \mathfrak{g}$ is the identity. Hence there is an open neighbourhood Ω of 0 in $\mathfrak{g}$ such that $U = \exp_G(\Omega)$ is open in G and $\exp_G : \Omega \to U$ is an analytic diffeomorphism.$(U, \{\exp_G^{-1} : U \to \Omega)$ is a coordinate chart at 1 for G.*

4.30. Proposition. *If k is non-archimedean, there is a $d > 0$ such that $\exp_G(B_{d'(0)})$ is an open compact sugroup of G for all $d' \leq d$.*

This proposition follows from 4.6 to 4.7 and the existence of a coordinate chart as above.

4.31. Remarks. (**1**) From the uniqueness of the solution for the initial value problem for ordinary differential equations, one concludes that the germ at 0 of the exponential map – a map satisfying (∗) and whose tangent map is the identity of $\mathfrak{g}$ – is unique.

(**2**) We use the notation in 4.25 (5). While the germ is unique, when k is not archimedean, there can be several distinct maps that have the same germ satisfying (∗) and with tangent map at 0 as identity. Here is an illustration of the phenomenon: let $G = \mathfrak{o}_k \times F_k$. Its Lie algebra is k. Set $\exp_G : \mathfrak{o}_k \to G$ to be the inclusion $x \to (x, 0)$ of $\mathfrak{o}_k$ in G. Define $\exp'_G$ as follows: $\exp'_G(x) = (x, \varpi_k(x))$. Then $\exp_G = \exp'_G$ on $\mathfrak{p}_k$ and hence they have the same germ at 0 but are not equal, as $\varpi_k(1) = 1(\in F_k)$. This is essentially a reformulation of 4.25 (5). Despite this non-uniqueness we will talk of *the* exponential map on $B_d(0)$ for some $d > 0$.

(**3**) When k is archimedean, as we saw above we can define $E_X(t)$ as an analytic function on all of $\mathfrak{g} \times k$. Hence $\exp_G(X)$ can be defined as $E_X(1)$ for all $X \in \mathfrak{g}$. Moreover in view of the principle of analytic continuation, the germ of $\exp_G$ determines $\exp_G$ uniquely on all of $\mathfrak{g}$.

(**4**) It follows from the definitions and the uniqueness theorem for differential equations that we have an open neighbourhood Ω of 0 in $\mathfrak{g}$ such that $\exp_{R_{k/\mathbb{Q}}} = \exp_G$ on Ω. In the archimedean case $\Omega = \mathfrak{g}$.

4.32. Proposition. *Assume that k is non-archimedean and $\exp_G$, d be as in 4.30. Then $\exp_G$ is injective on $B_d(0)$.*

4.33. Proof. Note that since k is non-archimedean, if $X \in B_d(0)$, $r \cdot X \in B_d(0)$ for all $r \in \mathbb{Z}$. Suppose that $\exp_G(X) = \exp_G(Y)$ for some $X, Y \in B_d(0)$. Then we have $\exp_G(n \cdot X) = (\exp_G(X))^n = (\exp_G(Y))^n = \exp_G(n \cdot Y)$ for all $n \in \mathbb{Z}$. Since $\mathbb{Z}$ is dense in $\mathbb{Z}_p$, we see that $\exp_G(t \cdot X) = \exp_G(t \cdot Y)$ for all $t \in \mathbb{Z}_p$. Now – as is easy to see – $\{\exp_{H=R_{k/\mathbb{Q}_p}G} = \exp_G\}$. It follows that the tangent vector X to $\{\exp_H(tX) | \ t \in Z_p\}$ equals Y. Hence the proposition.

4.34. About the exponential map in $GL(V)$. (**1**)In the case when $k = GL(V)$, V a finite dimensional vector space, we have seen that the Lie algebra has a natural identification with $\mathfrak{gl}(V)$. The germ of the exponential map $\exp_{GL(V)} : \mathfrak{gl}(V) \to GL(V)$ is then given by the (convergent) series

$$\sum_{r=0}^{\infty} X^r/r! \qquad (*)$$

in a neighbourhood of zero in $\mathfrak{gl}(V)$. This follows from the uniqueness of the germ of the 1-parameter group of the vector field determined by X. This is the reason for naming the map, 'exponential' map.

(**2**) If $k = \mathbb{R}$ or $\mathbb{C}$, the series converges absolutely on all of $\mathfrak{gl}(V)$. In the case when k is non-archimedean and $G = GL(V)$, the issue of the domain of absolute convergence of the series $(*)$, is more delicate. Recall that the spectral radius $\sigma(X)$ of $X \in \mathfrak{gl}(V)$ is the maximum of the absolute values of the eigenvalues of X (cf 1.96). As was observed in 1.97, $\sigma(X) = \mathrm{Lim}_{r \to \infty} ||X^r||^{1/r}$ and a necessary and sufficient condition for the absolute convergence of $(*)$, when k is a non-archimedean, is that $\sigma(X) < p^{-1/(p-1)}$, where p is the residue field characteristic of k. Set

$$\Omega_{GL(V)} = \{X \in \mathfrak{gl}(V)|\ \sigma(X) < p^{-1/(p-1)}\}.$$

Then $\Omega_{GL(V)}$ is an open subset of $GL(V)$ and an exponential map $\exp: \Omega_{GL(V)} \to GL(V)$ is given by the absolutely convergent series $(*)$.

4.35. Remarks. (**1**) Assume that k is non-archimedean. Then the map $t \rightsquigarrow \exp(t \cdot X)$ (with X sufficiently close to 0 in $\mathfrak{g}$) is an analytic group homomorphism of $\mathfrak{o}_k$ in G whose differential maps the tangent vector v in the tangent space at $t \in \mathfrak{o}_k$ ($\simeq k$) into $v \cdot X$ and is thus injective. It follows in the light of the above proposition that $(\mathfrak{o}_k, u_X : \mathfrak{o}_k \to G)$ is a closed sub-manifold of G: here $u_X(t) = \exp_G(t \cdot X)$.

(**2**) When k is archimedean, the map $k \ni t \rightsquigarrow \exp(t \cdot X)$ is an analytic homomorphism of k into G whose differential at $t \in k$ takes $(d/dt)_t$ to $X_{\exp(t \cdot X)}$. If $X \neq 0$ this implies that the map is injective on a neighbourhood of 0 in k. It follows that the kernel L of this analytic homomorphism is a closed discrete subgroup of k. We thus obtain an injective analytic homomorphism u_X of the Lie group k/L into G which is also an immersion (see 4.10 (9)). Clearly $(k/L, u_X : k/L \to G)$ is an analytic sub-manifold of G. In general $u_X(k)$ need not be closed in G.

4.36. Proposition. *Let $X, Y \in \mathfrak{g}$ be such that $[X, Y] = 0$. Then for $t, s \in k$ such that $\exp(t \cdot X)$ and $\exp(s \cdot Y)$ are defined, they commute, and $\exp(t \cdot X) \cdot \exp(s \cdot Y) = \exp(t \cdot X + s \cdot Y)$ (when both sides are defined). In particular, $\exp(t \cdot X) \cdot \exp(s \cdot X) = \exp((t + s) \cdot X)$.*

This is an immediate consequence of Corollary 3.47.

4.37. Corollary. *Let G be a Lie group such that $\mathfrak{g}$ is abelian. Then G admits an open subgroup G_o which is abelian.*

4.38. Proof. One need only take G_o to be the subgroup generated by $\exp_G(\Omega)$ for an open neighbourhood Ω of 0 in $\mathfrak{g}$. We note that when k is archimedean, G_o is the connected component of the identity in G.

The following proposition follows from Proposition 4.3 and the fact that the differential of the exponential map at 1 is the identity (4.29).

4.39. Proposition. *Let* $\mathfrak{g} = \coprod_{1 \leq i \leq r} \mathfrak{v}_i$ *be a direct sum decomposition of* $\mathfrak{g}$ *as a (finite) direct sum of vector subspaces* $\{\mathfrak{v}_i | i \in [1, r]\}$. *Let* $\delta > 0$ *be such that* $\exp_G$ *is defined on* $B_\delta(0)$. *Set* $\Omega_i = \mathfrak{v}_i \cap B_\delta(0)$ *and let* $\mathcal{E} : (\mathfrak{g} \supset) \Omega = \prod_{i \in [1, r]} \Omega_i \to G$ *be the map defined by setting for* $\underline{v} = \{v_i \in \mathfrak{v}_i | i \in [1, r]\} \in \coprod_{i \in [1, r]} \Omega_i$, $\mathcal{E}(\underline{v}) = \exp_G(v_1) \cdot \exp_G(v_2) \cdots \exp_G(v_r)$. *Then* $d\mathcal{E}_0 = Id_{\mathfrak{g}}$, *the identity map of* $\mathfrak{g}$ *on itself.*

4.40. Corollary. *There is an open neighbourood* $\Omega_{\mathcal{E}}$ *of* 0 *in* $\mathfrak{g}$ *such that* $\mathcal{E}$ *is defined on* $\Omega_{\mathcal{E}}$ *and is an analytic diffeomorphism of* $\Omega_{\mathcal{E}}$ *onto an open neighbourhood* $U_{\mathcal{E}}$ *of the identity in* G.

The result below is easily deduced from Corollary 4.40 and 4.6−4.7.

4.41. Theorem. *Let* $B = \{e_i | 1 \leq i \leq m\}$ *be a basis of* $\mathfrak{g}$. *Identify* $\mathfrak{g}$ *with* k^n *through the isomorphism given by the basis. The standard norm* $|| \cdot ||$ *on* k^m *defines a norm on* $\mathfrak{g}$ *through this isomorphism, which will also be denoted* $|| \cdot ||$. *For* $T \in End_k\mathfrak{g}$, *set* $||T|| = Sup\{||T(v)|| \mid v \in \mathfrak{g}, ||v|| = 1\}$. *For* $r > 0$ *let* $B_r(0) = \{v \in \mathfrak{g} | \, ||v|| < r\}$. *Then there exists,* $d_1 > 0, d_1 \leq d$ *such that the following holds:*
(i) *exp is defined on* $B_d(0)$ *and is an analytic diffeomorphism onto an open neighbourhood* U *of* 1 *in* G.
(ii) *For* $X, Y \in B_{d_1}(0)$, $\exp(X) \cdot \exp(Y) \subset U$ *and the Taylor series at* $(0, 0)$ *of the analytic function* $F = \exp^{-1} \circ \mu \circ (\exp \times \exp) : (B_{d_1}(0) \times B_{d_1}(0)) \to B_d(0)(\subset k^m)$ *converges uniformly and absolutely to the function on* $B_{d_1}(0) \times B_{d_1}(0)$.
(iii) *The series* $\sum_{r=0}^{\infty} ad(X)^r / r!$ *converges absolutely and uniformly on* $B_{d_1}(0)$ *to* $Ad_{GL(\mathfrak{g})} (\exp_G(X))$.
(iv) *If* k *is non-archimedean, for all* $d' \leq d_1$ *and* $U' = \exp(B_{d'}(0))$ *is a (compact open) subgroup of* G.

4.42. Definitions. An *exponential coordinate chart on* G is a coordinate chart at 1 of the form $(U = \exp_G(B_d(0)), \, \exp_G^{-1} : U \to B_d(0))$ where d is as in the theorem above.

A Lie group G over k to be of *exponential type* if it admits an exponential coordinate chart with the additional properties described below.

When k is archimedean G is generated by $\exp_G(B_d(0))$. This is equivalent to the stipulation that G is connected.

When k is non-archimedean, $G = u = \exp_G(B_d(0))$ and the Taylor series of the (vector valued) function $F = \exp_G^{-1} \circ \mu \circ (\exp_G \times \exp_G) : B_d(0) \times B_d(0) \to B_d(0)$ at $(0, 0)$ converges absolutely and uniformly F in $B_d(0) \times B_d(0)$. Further, for every $0 < d' \leq d$, $\exp_G(B_{d'}(0))$ is (a compact open) subgroup of G.

Note that any Lie group over k admits an open subgroup of exponential type (cf. 4.6–4.7). An exponential coordinate system as above will be called a *good coordinate system*.

4.43. Remarks. (**1**) When k is archimedean, the exponential map $\exp_G$ is uniquely defined on all of $\mathfrak{g}$. The connected component of the identity in G is the unique open subgroup of exponential type in G. It is generated by $\exp_G(\mathfrak{g})$.

(**2**) When k is non-archimedean and G is of exponential type, we will speak of 'the' exponential map; we mean by that the exponential map that enabled the definition of the group of exponential type. It is defined on $B_d(0)$ for some $d > 0$ with $G = \exp_G(B_d(0)) = G$ (and so G is the group generated by $\exp_G(B_d(0))$).

(**3**) Any Lie group G admits an open subgroup of exponential type. When k is archimedean, this open subgroup of exponential type is unique – it being the connected component of the identity. When k is non-archimedean, G admits a family of compact open subgroups of exponential type which constitute a fundamental system of neighbourhoods of 1 in G. If $(U = \exp_G(B_d(0)), \exp_G^{-1} : U \to B_d(0))$ is a coordinate system as in the definition, for any $d' \leq d$, $\exp_G(B_{d'}(0))$ is a Lie group of exponential type (cf. 4.6, 4.7, 4.41).

4.44. Theorem. *The assignment to a Lie group of its Lie algebra is 'functorial': let G_1, G_2 be Lie groups and $f : G_1 \to G_2$ be an analytic group homomorphism. Then $df_1 : T_1 G_1 \to T_1 G_2$ regarded as a map of the Lie algebra $\mathfrak{g}_1$ in $\mathfrak{g}_2$ is a Lie algebra homomorphism. Let $B_{d_2}(0)_2$ be an exponential neighbourhood of 0 in $\mathfrak{g}_2$ and $B_{d_1}(0)_1 \subset df_1^{-1} B_d(0)_2$ be an exponential neighbourhood of 0 in $\mathfrak{g}_1$. Then one has $\exp(df(X)) = f(\exp(X))$ for $X \in B_{d_1}(0)_1$. In other words the following diagram is commutative:*

$$B_{d_1}(0)_1 \overset{df_0}{\to} B_{d_2}(0)_2$$
$$\exp_{G_1} \downarrow \qquad \downarrow \exp_{G_2}$$
$$G_1 \overset{f}{\to} G_2$$

df_1, also denoted $\dot{f}$, is the Lie algebra morphism *induced by f.*

4.45. Proof. Since f is an analytic group homomorphism, one sees that $df_g (L_g(v)) = L_{f(g)}(df_1(v))$ for $v \in T_1 G_1$ (for the definition of L_g, see 4.11). Let $X, Y \in \mathfrak{g}_1$; then the 1-parameter group of X is $\{R_{\exp(-tX)}\}_{|t|<d}$ for a suitable d. It is immediate from this that the 1-parameter group of $df(X)$ is $\{R_{\exp(-tdf(X))}\}_{|t|<d}$. One then has $[X, Y]_1 = d/dt(R_{\exp(-tX)*}(Y_1))|_{t=0}$ (see 3.45). It follows that

$$df([X, Y]_1) = d/dt(R_{f(\exp(-tX)*)}(df_1(Y_1)))|_{t=0} = [df(X), df(Y)]_1$$

and hence $[df(X), df(Y)] = df([X, Y])$. The second assertion is immediate from the definition of the exponential map.

4.46. Representations. A(n *analytic*) *representation* of a Lie group G (over k) is a finite dimensional representation $\rho : G \to GL(V_\rho)$ of G as a group such that that ρ is an analytic map. The various definitions made in the context of representations of groups (see 1.64–1.65) carry over verbatim to representations of Lie groups: sub and quotient representations, direct sums of representations, $\mathrm{Hom}(\rho, \sigma)$ for two representations ρ and σ, dual of a representation, tensor product of representations etc. The analyticity requirement will be automatically satisfied when we construct new representations out of old so long as the old ones are analytic representations.

4.47. By Theorem 4.44, a representation $\rho : G \to GL(V)$ induces a representation $\dot{\rho}$ of the Lie algebra $\mathfrak{g}$ in $\mathfrak{gl}(V)$. Let ρ, ρ' be representations of G. A morphism $f : \rho \to \rho'$ is simply a morphism of abstract group representations. It is clear from the definitions that if f is a morphism of ρ in ρ', it is also a morphism of $\dot{\rho}$ in $\dot{\rho}'$. It is also clear that direct sums and tensor products of representations of a Lie group G induce the corresponding direct sums and tensor products of $\mathfrak{g}$, the Lie algebra of G. If ρ and σ are representations of the Lie group G, $\dot{\mathrm{Hom}}(\rho, \sigma)$ has a natural identification with $\mathrm{Hom}(\dot{\rho}, \dot{\sigma})$. If ρ is a (n analytic) representation of the Lie group G, we have the representation $\otimes^{p,q} \rho$ of G as an abstract group on the space of p-times covariant and q-times contravariant tensors. This again is an analytic map of G in $GL(\otimes^{p,q} V_\rho)$. Similarly the exterior powers $\bigwedge^p \rho$ on $\bigwedge^p V_\rho$ of ρ as representations of G as an abstract group are analytic representations. We summarize these observations in the following:

4.48. Theorem. *Let ρ and ρ' be representations of a Lie group G and $\dot{\rho}$ and $\dot{\rho}'$ the induced representations of its Lie algebra $\mathfrak{g}$. Then we have the following natural isomorphisms.*
(i) $\dot{\rho \oplus \rho'} \simeq \dot{\rho} \oplus \dot{\rho}'$,
(ii) $\dot{\rho \otimes \rho'} \simeq \dot{\rho} \otimes \dot{\rho}'$,
(iii) $\dot{\mathrm{Hom}}(\rho, \rho') \simeq \mathrm{Hom}(\dot{\rho}, \dot{\rho}')$,
(iv) $\dot{\rho}^* \simeq \dot{\rho}^*$,
(v) $\dot{\otimes}^{p,q} \rho \simeq \otimes^{p,q} \dot{\rho}$ *and*
(vi) $\dot{\bigwedge}^p \rho \simeq \bigwedge^p \dot{\rho}$.

4.49. Theorem. (i) *Let ρ be a representation of G. Then for a vector $v \in V_\rho$, if $\rho(g)(v) = v$ for all $g \in G$, $\dot{\rho}(X)(v) = 0$ for all $X \in \mathfrak{g}$. Conversely if $\dot{\rho}(X)(v) = 0$ for all $X \in \mathfrak{g}$, there is an open subgroup G' in G such that $\rho(g)(v) = v$ for all $g \in G'$. When k is archimedean, G' contains the connected component of the identity in G.*
(ii) *If σ is a sub-representation of a representation ρ of G, $\dot{\sigma}$ is a sub-representation of $\dot{\rho}$ of $\mathfrak{g}$ and $\dot{\rho/\sigma} = \dot{\rho}/\dot{\sigma}$. Conversely if $(\dot{\sigma}, W)$ is a sub-representation of $\dot{\rho}$, there is an open subgroup G' of G such that $\rho(G')$*

$(W) \subset W$. *When k is archimedean, G' may be taken as the connected component of the identity in G.*

4.50. Proof. The theorem follows from Theorem 4.44 and the fact that for $X \in \mathfrak{g}$ and $\dot{\rho}(X)$ is tangential to $\rho(\exp(t \cdot x))$, $t \in k$ near 0 so that $\dot{\rho}(X)(v) = d/dt|_{t=0} \rho(\exp(tX))(v)$. One also uses the explicit description of the exponential map in $GL(n, k)$ (see 4.34).

4.51. Corollary. *Let G be a Lie group and $\mathfrak{g}$ its Lie algebra. Let ρ be a finite dimensional representation of G. Let E be a k-linear subspace of V_ρ of dimension q and v a non-zero vector in $\bigwedge^q E \subset \bigwedge^q V_\rho$. Let $G_E = \{g \in G|\ \rho(g)(E) = E\}$ and $G_E^o = \{g \in G|\ \bigwedge^q \rho(g) \cdot v = v\}$. Then the Lie algebra $\mathfrak{g}_E$ of G_E (resp. $\mathfrak{g}_E^o$ of G_E^o) is $\{X \in \mathfrak{g}|\ \dot{\rho}(X)(E) \subset E\}$ (resp. $\{X \in \mathfrak{g}|\ \bigwedge^q \dot{\rho}(X)(v) = 0\}$.)*

4.52. The Adjoint Representation. For an element $g \in G$, we denote by $\mathrm{Int}(g) : G \to G$ the analytic group isomorphism defined by setting for $h \in G$, $\mathrm{Int}(g)(h) = g \cdot h \cdot g^{-1}$. Evidently, the map $(g, h) \rightsquigarrow \mathrm{Int}(g)(h)$ is an analytic map of $G \times G$ in G. Now since $\mathrm{Int}(g)$ is an analytic group isomorphism, it induces a Lie algebra automorphism $\dot{\mathrm{Int}}(g) : \mathfrak{g} \to \mathfrak{g}$. We denote $\dot{\mathrm{Int}}(g)$ by $Ad(g)$. Then Ad is an analytic homomorphism of G in $GL(\mathfrak{g})$; in other words Ad is a representation of G on the vector space $\mathfrak{g}$. It is the *adjoint* representation of G. On the other hand, we defined in 1.48, the adjoint representation $ad : \mathfrak{g} \to \mathfrak{gl}(\mathfrak{g})$ of the Lie algebra $\mathfrak{g}$. It is easy to see that we have:

4.53. Lemma. $ad = \dot{Ad}$. *Let ρ be a representation of G. Then given a compact subset B of G, there is a $d > 0$ such that for $X \in B_d(0)$ sufficiently near 0, $\rho(Ad(g)(\exp(X))) = \rho(g) \cdot (\exp(\dot{\rho}(X))) \cdot \rho(g)^{-1}$.*

Properties of representations $\dot{\rho}$ of $\mathfrak{g}$ have implications for ρ.

4.54. Proposition. *If ρ is a representation of G such that $\dot{\rho}$ is irreducible, then ρ is irreducible. Conversely if ρ is irreducible $\dot{\rho}$ is completely reducible.*

4.55. Proof. The first assertion holds since every $\rho(G)$-stable subspace of V_ρ is $\mathfrak{g}$-stable. Let $W \neq \{0\}$ be an irreducible $\mathfrak{g}$-submodule of V_ρ. Then for any $g \in G$, $\rho(g)(W)$ is $\mathfrak{g}$-stable (by the lemma above). It is easily seen to be an irreducible $\mathfrak{g}$-module. Now $\sum_{g \in G} \rho(g)(W)$ is G-stable and hence equal to V_ρ. Hence V_ρ as a module over $\mathfrak{g}$ is a sum of irreducible $\mathfrak{g}$-modules and is thus completely reducible (see 1.53).

4.56. Corollary. *If a representation ρ of G is completely reducible then $\dot{\rho}$ is completely reducible.*

4.57. Definition. Let ρ (resp. σ) be a finite dimensional representation of G (resp. $\mathfrak{g}$). Define the symmetric bilinear form B_ρ (resp. B_σ) on $\mathfrak{g}$ by setting for $X, Y \in \mathfrak{g}$, $B_\rho(X, Y) = \mathrm{Trace}(\dot{\rho}(X) \cdot \dot{\rho}(Y))$ (resp. $B_\sigma(X, Y) = \mathrm{Trace}(\sigma(X) \cdot \sigma(Y)))$. The *Killing form* $B_\mathfrak{g}$ is the bilinear form B_{ad}. For a representation ρ of a Lie group G with $\mathfrak{g}$ as its Lie algebra, $B_\rho = B_{\dot{\rho}}$.

4.58. Corollary. *For any representation ρ (resp. σ of $\mathfrak{g}$), the form B_ρ on $\mathfrak{g}$ is invariant under $Ad(G)$ (resp. $ad(\mathfrak{g})$), i.e.,*

for $g \in G$ and $X, Y \in \mathfrak{g}$, $B_\rho(Ad(g)(X), Ad(g)(Y)) = B_\rho(X, Y)$.
(resp. $B_\sigma([Z, X], Y) + B_\sigma(X, [Z, Y]) = 0$, $\forall X, Y, Z \in \mathfrak{g}$)

4.59. Proof. From Lemma 4.53, one deduces easily that for $g \in G$ and $X \in \mathfrak{g}$, $Ad(g) \cdot ad(X) \cdot Ad(g)^{-1} = ad(Ad(g)(X))$. The corollary now follows from the fact that the trace of a linear endomorphism of a vector space is invariant under conjugation by linear automorphisms. The statement for σ is a consequence of the fact that $\mathrm{trace}(A \cdot B) = \mathrm{trace}(B \cdot A)$ for any two endomorphisms A, B of a vector space.

4.60. Definition. A *Lie subgroup* of G is an analytic) sub-manifold $(H, i : H \to G)$ (see 3.25) with H a Lie group and i an (injective analytic) group homomorphism.

With this definition we have the following Corollary to Lemma 4.53.

4.61. Corollary. *Let G be Lie group and $(H, i : H \to G)$ a Lie subgroup. If $i(H)$ is normal in G, $\mathfrak{h}$ is an ideal in $\mathfrak{g}$.*

4.62. Examples of Lie Subgroups. (**1**) Let k be non-archimedean. Let B be a compact open $\mathfrak{o}_k$-submodule on which the exponential map is defined. For $X \in B$, let $E_X : \mathfrak{o}_k \to G$ be the map defined by setting for $t \in \mathfrak{o}_k$, $E_X(t) = \exp(t \cdot X)$. Then if $X \neq 0$, $(\mathfrak{o}_k, E_X)$ is a Lie subgroup of G (see 4.25 (1)).

(**2**) When k is archimedean, define for $X \in \mathfrak{g}$, $E_X : k \to G$ by setting for $t \in k$, $E_X(t) = \exp(t \cdot X)$. Then E_X is an analytic group homomorphism. Let L be the kernel of this homomorphism. If $X \neq 0$, L is a closed discrete subgroup of k and $H = k/L$ is a Lie group of dimension 1. The map E_X factors through H, giving an injective analytic homomorphism $\bar{E}_X : H \to G$ making $(H, \bar{E}_X)$ into a Lie subgroup of G.

(**3**) One of the examples exhibited in Chapter 3, of a sub-manifold whose manifold topology is not the one induced from the ambient manifold (see 3.26 (1)), is a special case of (2) above. The ambient Lie group G is $\mathbb{S}^1 \times \mathbb{S}^1$. The Lie algebra of G is isomorphic to $\mathbb{R} \times \mathbb{R}$ under an isomorphism for which the exponential map $\exp_G : \mathbb{R} \times \mathbb{R} \to \mathbb{S}^1 \times \mathbb{S}^1$ takes $(t_1, t_2) \in \mathbb{R} \times \mathbb{R}$ to $(e^{2\pi i t_1}, e^{2\pi i t_2})$.

Let $u : \mathbb{R} \to G$ defined as follows: let α be an irrational number and for $t \in \mathbb{R}$, set $u(t) = (e^{2\pi i t}, e^{2\pi i \alpha t})$. evidently, $\mathbb{R}, u : \mathbb{R} \to G$ is a Lie subgroup.

(**4**) On the other hand, in most natural examples of Lie subgroups, the topology induced from the ambient group is the same as the manifold topology on the subgroup. Clearly the inclusion i of k^r in k^m for $r \leq m$ makes $(k^r, i : k^r \to k^m)$ an analytic sub-manifold of k^m. The inclusions of $GL(r, k), SL(r, k), SO(r)$ in $GL(n, k), n \geq r$ provide other examples. In all these examples the topology induced on the Lie subgroup by the topology of the ambient group is the manifold topology on the sub-manifold. All these are examples of the following phenomenon.

4.63. Proposition. *Let $(H, i : H \to G)$ be a Lie subgroup of the Lie group G. If $i(H)$ is a closed subset of G, the manifold topology on H is the same as the one induced on it by G through i.*

4.64. Proof. Let U be an open neighbourhood of 1 in H and V a compact neighbourhood of 1 such that $V^{-1} \cdot V \subset U$. Let D be a countable dense subset of H such that $H = D \cdot V$. Now $i(V)$ is a compact, hence closed, subset of $i(H)$. $i(H)$ being a closed subset of G, it is locally compact in the topology induced from G and hence the Baire Category Theorem (1.73) holds in $i(H)$. It follows that since $i(D) \cdot i(V) = i(H)$, $i(V)$ has a non-void interior $i(V)^o$ in $i(H)$. If $x \in i(V)^o$, $x^{-1} \cdot i(V) (\subset i(W))$ is a neighbourhood of 1 in $i(H)$. It follows that a neighbourhood of 1 in H maps onto a neighbourhood of 1 in $i(H)$. Since i is a group homomorphism, we see that a neighbourhood of any $h \in H$ maps onto a neighbourhood of $i(h)$ in $i(H)$. Thus the map i is a homeomorphism of H on $i(H)$. Hence the proposition.

4.65. Theorem. *Let G be a Lie group and $(H, i : H \to G)$ be a Lie subgroup. Assume that $i(H)$ is closed in G. Then the coset space $G/i(H)$ carries a natural structure of an analytic manifold such that the underlying manifold topology on $G/i(H)$ is the quotient topology on $G/i(H)$. The map $\mu' : G \times G/i(H) \to G/i(H)$ defined by setting for $(x, y \cdot i(H)) \in G \times G/i(H)$, $\mu'(x, y \cdot i(H)) = x \cdot y \cdot i(H)$, is analytic. If $i(H)$ is normal in G, μ' factors through to the multiplication on $G/i(H)$.*

4.66. Proof. We identify H with its image $i(H)$ and i with the inclusion of H in G. The Lie algebra $\mathfrak{h}$ of H is then identified as a subalgebra of $\mathfrak{g}$. Let $\mathfrak{v}$ be a vector subspace of $\mathfrak{g}$ such that $\mathfrak{g} = \mathfrak{v} \oplus \mathfrak{h}$. Then there are open neighbourhoods W of 0 in $\mathfrak{h}$ and V of 0 in $\mathfrak{v}$ such that the map $\mathcal{E} : V \times W \to G$ defined by setting for $(v, w) \in V \times W$, $\mathcal{E}(v, w) = \exp(v) \cdot \exp(w)$ is an analytic diffeomorphism of $V \times W$ onto an open subset Ω of G (see Corollary 4.40). We equip G/H with the quotient topology. Let $\pi : G \to G/H$ be the natural map. Then since $\exp(V') \cdot H$ is an open saturated set for every open subset V' in V we conclude that $\Psi = \pi \circ \exp$ is a homeomorphism of V on an open set U of G/H. Let $\Phi : U \to V$ be the inverse of Ψ. Let $\underline{A} = \{(g \cdot U, \Phi \circ L_g^{-1}) | g \in G\}$

where L_g denotes the action (on the left) by $g \in G$ on G/H. It is now not difficult to see that $\underline{A}$ is an analytic atlas on G/H and the analytic manifold structure it defines on G/H has the properties required in the theorem. The proof of the last assertion is left to the reader.

We can now establish

4.67. Theorem. *Assume that k is archimedean. Let G be a connected abelian Lie group of dimension n over k. Then there is a (finite) set $S = \{e_i \mid 1 \leq i \leq r\}$ of vectors in k^n linearly independent over $\mathbb{R}$ and an analytic group isomorphism $f : k^n/L \simeq G$ where L is the (free abelian) subgroup of k^n generated by S. G is a divisible group.*

4.68. Proof. The last assertion follows from the fact that k^n is a divisible group (and hence so is any quotient). We know from proposition 4.36 that $\exp : \mathfrak{g} \to G$ is an analytic group homomorphism of $\mathfrak{g}$ $(\simeq k^n)$ into G. Since $d(\exp)_e$ is an isomorphism and $\exp$ is a group homomorphism, $d(\exp_v)$ is an isomorphism at all $v \in \mathfrak{g}$. It follows (by the inverse function theorem) that $\exp(\mathfrak{g})$ is an open (hence also closed) subgroup of G. It follows that $\exp(\mathfrak{g}) = G$. If L is the kernel of $\exp$, by 4.65, $\exp$ descends to an isomorphism of the group $\mathfrak{g}/L$ on G. It remains to show that L is generated as an abstract group by a set of vectors linearly independent over $\mathbb{R}$. This follows from the following.

4.69. Proposition. *Let L be a closed subgroup of $\mathbb{R}^n$ such that $L \cap U = \{0\}$ for some neighbourhood U of 0 in $\mathbb{R}^n$. Then L is generated as an abstract group by a subset $S \subset L$ of vectors in $\mathbb{R}^n$ linearly independent over $\mathbb{R}$.*

4.70. Proof. We argue by induction on n. Let V be the $\mathbb{R}$-linear span of L. If $\dim_{\mathbb{R}} V < n$, the proposition holds by induction hypothesis. Hence we may assume that $V = \mathbb{R}^n$. It follows that there exists a basis $B = \{e_i \mid 1 \leq i \leq n\}$ of $\mathbb{R}^n$ contained in L. Let L' be the subgroup of L generated by B. Then one has an isomorphism of $\mathbb{R}^n/L' \simeq (\mathbb{S}^1)^n \overset{\text{def}}{=} T^n$ so that we get an analytic group homomorphism $\pi : \mathbb{R}^n \to (\mathbb{S}^1)^n$. One sees from the fact that $L \cap U = \{0\}$, that $\pi(L)$ is discrete in the topology induced on it from the compact group T^n and is therefore finite. It follows that L is finitely generated and as it is torsion-free it is a free abelian group on a subset S of L. Since $L/L'(= \pi(L))$ is finite, the rank of L is the same as that of L'; and as L' spans $\mathbb{R}$ as a $\mathbb{R}$-vector space, so does L. Thus S contains a basis S_o of $\mathbb{R}^n$. Arguing as above with L_o, the subgroup of k^n generated by S_o replacing L', we see that L/L_o is finite. It follows that $S = S_o$ proving the theorem.

The proofs of the following two results which are easily deduced from Proposition 4.69 are left as exercises to the reader.

4.71. Corollary to Theorem 4.67. *Let G be an abelian Lie group over an archimedean field k. Let G^o be the identity connected component of G. Then $G \simeq G^o \times (G/G^o)$.*

4.72. Proposition. *Let H be a closed subgroup of $\mathbb{R}^n$. Then there are subsets S_1, S_2 of H such that $S_1 \cup S_2$ is linearly independent over $\mathbb{R}$ and $H = \coprod_{s \in S_1} \mathbb{R} \cdot s \oplus \coprod_{s \in S_2} \mathbb{Z} \cdot s$.*

Proposition 4.72 is equivalent to the following well-known result.

4.73. Theorem (L Kronecker). *Let $a = \{a_i\}_{1 \leq i \leq n} \in \mathbb{R}^n$. Then the set $\{m \cdot \alpha + \mathbb{Z}^n | m \in Z\}$ is dense in $\mathbb{R}^n$ if and only if $\{a_i | 1 \leq i \leq n\}$ are linearly independent over $\mathbb{Q}$.*

4.74. Theorem. *A compact connected complex Lie group G is abelian.*

4.75. Proof. Let $(U, f = \{f_i | 1 \leq i \leq m\} : U \to \Omega \subset \mathbb{C}^m)$ be a coordinate chart at 1. Then given $g \in G$ there is an open neighbourhood B_g of 1 in G and an open neighbourhood V_g of g such that $h \cdot B_G \cdot h^{-1} \subset U, \ \forall \ h \in V_g$. Now since G is compact and $\{V_g | \ g \in G\}$ is an open covering of G, there is a finite subset S of G such that $\cup_{g \in S} V_g = G$. Set $B = \cap_{g \in S} B_g$; then B is a compact neighbourhood of 1 in G. For an element $u \in B$ and $1 \leq i \leq m$, define the map $\Phi_{iu} : G \to \mathbb{C}$ by setting for $h \in G$, $\Phi_{iu}(h) = f_i(h \cdot u \cdot h^{-1})$. The map is well defined since every $h \in G$ is in V_g for some $g \in S$. As Φ_{iu} is continuous, and G compact, $|\Phi_{iu}|$ attains its maximum at some point $h_o \in G$. Let $(W, \Theta : W \to B_1(0))$ be a chart at h_o. Then $\Phi_{ui} \circ \Theta^{-1}$ is a complex analytic function on the unit open disc in $\mathbb{C}^m$ whose absolute value takes its maximum at the interior point 0. By the maximum principle, Φ_{iu} is constant on the open subset W. By the principle of analytic continuation it is constant on all of G equal to $0 = \Phi_{iu}(1)$, i.e., $[h, u] = 1$ for all $h \in G$, $u \in B$. As B generates G, G is abelian.

4.76. Corollary. *Let G be a connected compact complex Lie group of dimension m over $\mathbb{C}$. Then there are $2 \cdot m$ vectors $\{v_i | 1 \leq i \leq 2 \cdot m\}$ in $\mathbb{C}^m$, linearly independent over $\mathbb{R}$ and a complex analytic group isomorphism $G \simeq \mathbb{C}^m / (\coprod_{i=1}^{2 \cdot m}) \mathbb{Z} \cdot v_i$.*

This follows from the above Theorem and Theorem 4.67.

4.77. Theorem. *Assume that k is non-archimedean. Let G be an abelian Lie group over k of dimension d. Let T be the torsion subgroup of G. Then T is a closed discrete subgroup of G. Let $L = \{x \in G | x^{p^n} \to 0 \text{ as } n \to \infty\}$. Then L is an open and closed subgroup of G isomorphic (as a Lie group) to an open subgroup of (the additive) group k^d. The map $\psi : T \times L \to G$ defined by setting for $(t, l) \in T \times L$, $\psi(t, l) = t \cdot l$, is an analytic diffeomorphism of $T \times L$ onto an open (hence closed) subgroup G' of G. The group G/G' is discrete and torsion-free.*

4.78. Proof. As G is abelian we will denote its group operation by '+'. This means that for $x \in G$ and $n \in \mathbb{Z}$, $n \cdot x$ denotes the nth power of x. Since G is abelian, it admits an open subgroup $U \simeq \mathfrak{o}_k^d$ where $d = \dim.G$. Now $p^n \cdot x \to 0$ as $n \to \infty$ for every $x \in k$. Thus $U \subset L$. Let $\varphi : U \to \mathfrak{o}_k^d (\subset k^d)$ be an isomorphism of U on $\mathfrak{o}_k^d$. Then φ extends uniquely to an injective morphism $\tilde{\varphi} : L \to k^d$ giving an isomorphism of L onto an open subgroup B of k^d. This is seen as follows: let $x \in L$. Then there is a unique minimal integer $r \geq 0$ such that $p^r \cdot x \in U$. Let $y \in k^d$ be the unique element in k^d such that $\varphi(p^r \cdot x) = p^r \cdot y$; define $\tilde{\varphi}(x) = y$. It is easy to see that $\tilde{\varphi}$ gives an isomorphism of L on an open subgroup of k^d. It follows that L is torsion-free so that $L \cap T = \{0\}$; and hence the map $\psi : T \times L \to G$ is an analytic diffeomorphism onto an open subgroup G' of G. Hence the topology on G/G' is discrete. It remains for us to prove that G/G' is torsion-free. If G/G' has an element of order $n > 1$, then it has an element x of prime order q, say. Let $y \in G$ be an element that maps to x under the natural map $G \to G/G'$. Then $q \cdot y = (u, v) \in T \times L = G'$. Since $u \in T$, $\exists m \in \mathbb{N}$ such that $m \cdot u = 0$. This implies that $m \cdot q \cdot y = m \cdot q \cdot v \in L$. In other words $p^l \cdot m \cdot q \cdot v \in U$ for some $l \in \mathbb{N}$, $r > 1$. Writing $p^l \cdot q \cdot m = b \cdot p^r$ with $(b, p) = 1$ and using the fact that $z \rightsquigarrow b \cdot z$ is an isomorphism of U on itself we conclude that $u \in L$. It follows that x, the image of (u, v) is 0, a contradiction to our assumption that x is of prime order. Thus G/G' is torsion-free.

5. Lie Groups: The Theorems of Cartan and Lie

5.1. The Lie algebra of a Lie subgroup, as we saw, is in a natural fashion a Lie subalgebra of the Lie algebra of the ambient group. A central result of Lie theory is a converse of this assertion, due to Lie and is known as the Fundamental Theorem of Lie Theory. Before proceeding to formulate and prove such a converse, we will establish a Lemma and a theorem due to Cartan, which we will use in the proof of the converse.

5.2. We retain the notation introduced in the last chapter. The next lemma is a useful characterization of the Lie subalgebra corresponding to a Lie subgroup of a Lie group.

5.3. Lemma. *Let G be a Lie group and $(H, i : H \to G)$ be a Lie subgroup of G. Then the Lie subalgebra $\mathfrak{h}$ of $\mathfrak{g}$ corresponding to H is*

$$\{X \in \mathfrak{g} | \ \exists \ \delta > 0 \ with \{\exp_G(I_\delta(0) \cdot X) \subset i(H)\}.$$

If k is archimedean and H connected, $i(H)$ is generated by $\exp_G(\mathfrak{h})$.

5.4. Proof. Let Ω'' be a relatively compact open neighbourhood of 0 in $\mathfrak{g}$ such that $\exp_G$ is an analytic diffeomorphism of Ω'' on an open set $U'' \ni 1$. Let Ω' be a relatively compact open subset of Ω'' with $0 \in \Omega'$. Set $U' = \exp_G(\Omega')$. Let $\Omega \ni 0$ be an open set of $\mathfrak{g}$ such that
(i) $-X \in \Omega$, $\forall \ X \in \Omega$,
(ii) if $U = \exp_G(\Omega)$, $U \cdot U \subset U'$ and
(iii) $\Phi' = di_1^{-1}(\mathfrak{h} \cap \Omega')$ is an exponential neighbourhood of 0 in $\mathfrak{h}$.
Set $\Phi = di_1^{-1}(\Omega)$, $V' = \exp_H(\Phi')$ and $V = \exp_H(\Phi)$. Let Γ be a countable subset of H with $V \cdot \Gamma = H$. Let $Y \in \mathfrak{g}$ be such that $\exp_G(t \cdot Y) \in i(H)$ for all $t \in I_\delta(0)$ for some $\delta > 0$. Then one has for every $t \in I_\delta(0)$, $\exp_G(t \cdot Y) = i(\alpha(t)) \cdot \gamma(t)$ with $\gamma(t) \in \Gamma$ and $\alpha(t) \in V$. As Γ is countable while $I_\delta(0)$ is not, there are infinitely many mutually distinct $\{\tau_n \in I_\delta(0) | \ n \in \mathbb{N}\}$ such that $\gamma(\tau_n) = \gamma$ (independent of n). Now $\exp_G(\tau_{n+1} - \tau_n) \cdot Y = i(\alpha(t_{n+1}) \cdot \alpha(t_n)^{-1})$ $= i(\exp_H(Z_n)) = \exp_G(di_1(Z_n))$ with $Z_n \in \Phi'$. Since $\exp_G$ is an ananalytic diffeomorphism on Ω' we conclude that $(\tau_{n+1} - \tau_n) \cdot Y = Z_n \in \mathfrak{h}$ proving the lemma.

The following result is due to E. Cartan when k is archimedean. The proof in the non-archimedean case is similar to Cartan's in the archimedean case; it is in fact easier.

5.5. Theorem (E. Cartan). *Let k be such that $\mathbb{Q}$ is dense in k. Let G be a Lie group over k and H a closed subgroup. Then H has a natural structure of an analytic manifold such that $(H, i : H \to G)$ where i is the inclusion of H in G, is a Lie subgroup of G (over k). Further the manifold topology on H is the same as that induced from G through i.*

5.6. Remarks. (**1**) If $\mathbb{Q}$ is dense in k, then $k \simeq \mathbb{Q}_p$ for some prime p or $k \simeq \mathbb{R}$ according as k is non-archimedean or archimedean.

(**2**) Suppose that k is *any* local field. Then as we have seen the Lie group G over k may be regarded as a Lie group over $\hat{\mathbb{Q}}$. Observe that if the $\hat{\mathbb{Q}}$ Lie algebra $\mathfrak{h}$ of H is a k-linear subspace of $\mathfrak{g}$, H is a Lie subgroup over k. This is seen as follows. The exponential map of $\mathfrak{h}$ in H (over $\hat{\mathbb{Q}}$) is the restriction of the exponential map of $\mathfrak{g}$ in G as a Lie group over $\hat{\mathbb{Q}}$. But this is the same as the exponential map when G is treated as a Lie group over k (4.31 (4)). It follows that $\exp_H$ considered as a map into G is an analytic map over k. Thus in a suitable open neighbourhood V of 1 in $\mathfrak{h}$, $(V, (\exp_H^{-1}(= \exp_G^{-1}|_V) : V \to W \subset \mathfrak{h}))$ is a coordinate chart. It is now easy to see that

$$\{(L_h(V), (\exp_H^{-1} \circ L_h^{-1})) : L_h(V) \to W)| \ h \in H\}$$

provides an analytic atlas on H (over k) making it into a Lie group over k with $R_{k/\hat{\mathbb{Q}}}H$ being the $\hat{\mathbb{Q}}$-structure we started off with.

5.7. Proof of 5.5 in the Case of Non-Archimedean k. Let Ω be a disc $B_d(0)$ in $\mathfrak{g}$ (identified with k^m through a k basis of $\mathfrak{g}$) on which the exponential map is defined and $(U = \exp(\Omega), \exp^{-1} : \exp(\Omega) \to \Omega)$ is a chart. Let $g \in U \cap H$. Then $g = \exp(X)$ for some $X \in \Omega$. Now $\exp(n \cdot X) = \exp(X)^n = g^n \in H$ for all $n \in \mathbb{Z}$. As $\mathbb{Q}$ is dense in k, the closure of $\mathbb{Z}$ in k is $I_1(0) = \{t \in k| \ |t| \leq 1\}$. Since H is closed in G, we conclude that $\exp(I_1(0) \cdot X) \in H$. Let $\mathfrak{h}$ be the k-span of $S = \{X \in \Omega| \ \exp(X) \in H\}$ and let $\{X_j| \ 1 \leq j \leq r\} \subset S$ be basis of $\mathfrak{h}$ over k. Let W be a subspace of $\mathfrak{g}$ such that $\mathfrak{g}$ is the direct sum of $\mathfrak{h}$ and W. Let

$$\mathcal{E} : \Omega_o = (\Omega \cap W) \times \prod_{j=1}^{r} (\Omega \cap (k \cdot X_j)) \to G$$

be defined by setting for $(w, \{t_j \cdot X_j\}_{0 \leq j \leq r}) \in \Omega_o$,

$$\mathcal{E}(w, \{t_j \cdot X_j\}_{0 \leq j \leq r}) = \exp(w) \cdot \prod_{j=0}^{r} \exp(t_j \cdot X_j).$$

Then $\mathcal{E}$ is an analytic diffeomorphism of an open neighbourhood Ω' of 0 in the product $W \times \prod_{j=0}^{r} k \cdot X_j$ of the form $B \times \prod_{j=0}^{r} I_\delta(0) \cdot X_j$, $B = B_c(0) \cap W$, for suitable $c, \delta > 0$, onto an open subset U' contained in U (Corollary 4.40). Suppose now that $g \in U' \cap H$, then $g = u \cdot h$ with $u \in \exp(B)$ and $h = \prod_{j=0}^{r} \exp(t_j \cdot X_j)$ with $t_j \in I_\delta(0)$. This means that $h \in H$ and hence $u \in H \cap U$. Now $u = \exp(Y)$ with $Y \in W$; on the other hand, as $u \in H \cap U$, $Y \in S$ and hence in the span $\mathfrak{h}$ of S. Since $W \cap \mathfrak{h} = \{0\}$, we conclude that $u = 1$. In other words, $\mathcal{E}^{-1}(H \cap U')$ is an open set in $\mathfrak{h} = \coprod_{j=0}^{r} k \cdot X_j$. It follows easily from this that H is an analytic sub-manifold and hence a Lie subgroup, the manifold topology being the same as the topology induced from G.

5.8. Proof of 5.5 in the Archimedean Case. As $\mathbb{Q}$ is dense in k, we have necessarily $k \simeq \mathbb{R}$. Let $S \subset \mathfrak{g}$ be the subset of those $X \in \mathfrak{g}$ for which there are sequences $\{X_n \in \mathfrak{g} | \ n \in \mathbb{N}\}$ tending to X and $\{t_n \in \mathbb{R} | \ n \in \mathbb{N}\}$ tending to 0, such that $\exp(t_n \cdot X_n) \in H$. Then we claim that $\exp(t \cdot X) \in H$ for all $t \in \mathbb{R}$. This is seen as follows. Let $X \in S$ and $\{X_n \in \mathfrak{g} | \ n \in \mathbb{N}\}$ and $\{t_n \in \mathbb{R} | n \in \mathbb{N}\}$ be sequences with $\exp(t_n \cdot X_n) \in H$ with X_n tending to X and t_n tending to 0. Now for $t \in \mathbb{R}$ and $n \in \mathbb{N}$, $g_n = \exp([t/t_n] \cdot t_n \cdot X_n) = (\exp(t_n \cdot X_n))^{[t/t_n]} \in H$ ($[t/t_n]$ is the largest integer $\leq t/t_n$). As g_n tends to $\exp(t \cdot X)$ and H is closed in G, $\exp(t \cdot X) \in H$, proving our contention. Let $\mathfrak{h}$ be the $\mathbb{R}$-linear span of S and $\{X_i | \ 1 \leq i \leq r\}$ a basis of $\mathfrak{h}$ contained in S. Let W be a vector subspace of $\mathfrak{g}$ such that $\mathfrak{g} = W \oplus \mathfrak{h}$. Let $\mathcal{E} : \mathfrak{g} \to G$ be the map which takes $Y + \sum_{j=0}^{r} t_j \cdot X_j$ ($Y \in W, t_j \in \mathbb{R}$) to $\exp(Y) \cdot \prod_{j=0}^{r} \exp(t_j \cdot X_j)$. Then evidently $\mathcal{E}(\{0\} \times \mathfrak{h}) \subset H$. Now $\mathcal{E}$ is an analytic diffeomorphism of an open set of the form $B \times V$ (where $B \ni 0$ is open in W and $V \ni 0$ is open in $\mathfrak{h}$) onto an open neighbourhood of 1 in G (Corollary 4.40). It suffices to prove that there is a neighbourhood Ω of 0 in $\mathfrak{h}$ such that $\mathcal{E}$ is an analytic diffeomorphism of Ω onto an open subset $U = \mathcal{E}(\Omega)$ and $U \cap H = U \cap \mathcal{E}(\mathfrak{h})$. Indeed if this is not the case we will have a sequence $\{g_n \in H\}_{n \in \mathbb{N}}$ tending to the identity in G with $g_n = \exp(Y_n) \cdot h_n$ where $Y_n \in W$ is a sequence tending to zero in W and $h_n \in H$. It follows that $\exp(Y_n) \in H$. Let $t_n = ||Y_n||$ and $X_n = Y_n / ||Y_n||$ ($||.||$ is some norm on $\mathfrak{g}$). Passing to a sub-sequence we may assume that X_n tends to a non-zero limit $X \in V$. On the other hand, $X \in S$ and hence belongs to $\mathfrak{h}$, a contradiction. That the manifold topology is the topology induced from G is proved as in the non-archimedean case. Hence the theorem (in the archimedean case).

5.9. Remark. That the examples (5)–(8) given in 4.10 are Lie groups follows also from Cartan's Theorem (5.5) together with Remarks in 5.6.

5.10. Corollary. *Let $\mathfrak{g}$ be a Lie algebra and $A(\mathfrak{g})$ be the group of all Lie algebra automorphisms of $\mathfrak{g}$. Then there is a Lie group structure over k on $A(\mathfrak{g})$ such that the map $A(\mathfrak{g}) \times \mathfrak{g} \to \mathfrak{g}$ given by $A(\mathfrak{g}) \times \mathfrak{g} \ni (T, X) \rightsquigarrow T(X)$ is analytic.*

5.11. Proof. $A(\mathfrak{g})$ is a closed subgroup of $GL(\mathfrak{g})$. It is therefore a Lie subgroup of $GL(\mathfrak{g})$ over $\hat{\mathbb{Q}}$ (Cartan's theorem). By 5.6 (2), it suffices to show that the Lie subalgebra corresponding to $A(\mathfrak{g})$ (over $\hat{\mathbb{Q}}$) is

$$\mathfrak{a}(\mathfrak{g}) = \{D \in \mathfrak{gl}(\mathfrak{g}) | \ D([Y, Z]) = [D(Y), Z] + [Y, D(Z)]\},$$

the Lie algebra of all derivations of $\mathfrak{g}$ (cf. 1.44) which is a k-linear subspace of $\mathfrak{gl}(\mathfrak{g})$. By 5.6 (2), this follows from the following:

Assertion *Let $X \in \mathfrak{gl}(\mathfrak{g})$. Then $X \in \mathfrak{a}(\mathfrak{g})$ if and only if the following holds: there is $\delta > 0$ such that for all $Y, Z \in \mathfrak{g}$ and all $t \in \hat{\mathbb{Q}}$ with $|t|_{\hat{\mathbb{Q}}} \le \delta$, we have:*

$$\exp_{GL(\mathfrak{g})}(t \cdot X)([Y, Z]) = [\exp_{GL(\mathfrak{g})}(t \cdot X)(Y), \exp_{GL(\mathfrak{g})}(t \cdot X)(Z)]$$

We leave the proof of the assertion as an exercise to the reader.

5.12. Corollary. *Assume that $\mathfrak{g}$ is the Lie algebra of a Lie group G of exponential type. Let $A(G)$ be the group of all analytic group automorphisms of G. Then $A(G)$ has a Lie group structure over $\hat{\mathbb{Q}}$ making it a Lie subgroup (over $\hat{\mathbb{Q}}$) of $GL_{\hat{\mathbb{Q}}}(\mathfrak{g})$. When k is non-archimedean, $A(G)$ is an open subgroup of $A(\mathfrak{g})$ and is thus a Lie group over k.*

5.13. Proof. The first assertion follows from Cartan's theorem. When k is non-archimedean, there is a $d > 0$ such that $\exp_G$ is a diffeomorphism *of* $B_d(0)$ on G. Let u be an automorphism G and $\dot{u}$ the induced automorphism of $\mathfrak{g}$. As G is of exponential type, $u(\exp_G(X)) = \exp_G(\dot{u}(X))$ for all $X \in B_d(0)$. The map $u \rightsquigarrow \dot{u}$ of $A(G)$ in $A(\mathfrak{g})$ is analytic and injective as is easily seen. Now let $A' = \{\alpha \in A(\mathfrak{g}) | \ \alpha(B_d(0)) = B_d(0)\}$. If $\{e_i\}_{1 \le i \le m}$ is the basis of $\mathfrak{g}$ defining the norm on $\mathfrak{g}$, $B_d(0) = \lambda \cdot (\sum_{i=1}^{m} \mathfrak{o}_k \cdot e_i)$ where $\lambda \in k$ with $|\lambda|_k = d$. It follows that A' is a compact open subgroup of $A(\mathfrak{g})$. Then $\tilde{\alpha} = \exp_G \circ \alpha \circ \exp_G^{-1}$ is an automorphism of G. The map $\alpha \rightsquigarrow \tilde{\alpha}$ is the inverse of the map $u \rightsquigarrow \dot{u}$ defined above. Thus $u \rightsquigarrow \dot{u}$ yields an isomorphism of $A(G)$ on A'. The latter is an open compact subgroup of the Lie group $A(\mathfrak{g})$ over k. Thus $A(G)$ is a Lie group over k. It is easy to see that the map $(u, g) \rightsquigarrow u(g)$ of $A(G) \times G$ in G is analytic over k.

The stronger result that $A(G)$ has a Lie group structure over k holds in the archimedean case as well and will be proved later.

5.14. Corollary. *Assume that $\hat{\mathbb{Q}} = k$. Let G, H be Lie groups over k. Then any continuous homomorphism f of G in H is analytic.*

5.15. Proof. Let $G_f = \{(g, h) \in G \times H | \ h = f(g)\}$, the graph of f. Then G_f is a closed subgroup of the Lie group $G \times H$ (over k) and hence is a Lie subgroup. The restriction $\pi = p_G|_{G_f}$ of the Cartesian projection p_G of $G \times H$ on G, to G_f is a bijective analytic group homomorphism. It follows that the rank $d\pi_e = \text{rank } d\pi_g$ for all $g \in G_f$. If this rank is less than $\dim_{\mathbb{R}} G$, the Implicit Function Theorem would imply that there is a compact neighbourhood U of e in G_f which maps into a proper analytic sub-manifold of G. This means that the compact set $\pi(U)$ has a void interior. Now, let $\{g_n \in G_f | \ n \in \mathbb{N}\}$ be a countable set such that $\bigcup_{n \in \mathbb{N}} g_n \cdot U = G_f$; then

$\forall\ n \in \mathbb{N}$, $\pi(g_n \cdot U) = \pi(g_n) \cdot \pi(U)$ has void interior. By the Baire category theorem (1.73), $G = \bigcup_{n \in N}\ \pi(g_n) \cdot \pi(U)$ has void interior, a contradiction. Thus $d\pi$ is surjective at all points of G_f. If $\dim_{\mathbb{R}} G_f > \dim_{.\mathbb{R}} G$, once again the Implicit Function Theorem would imply that π cannot be injective. Thus we see that $d\pi_g$ is an isomorphism for all $g \in G_f$. The Inverse Function Theorem implies that π^{-1} is analytic; thus π is an analytic diffeomorphism. Now $f = p_H|_{G_f} \circ \pi^{-1}$ where p_H is the Cartesian projection of $G \times H$ on H; and p_H being analytic, so is $p_H|_{G_f}$. Thus f is analytic.

5.16. Lemma. *Let G be a Lie group over k and let G' denote G considered as a Lie group over $k_o = \hat{\mathbb{Q}}$. Then the Lie algebra $\mathfrak{g}'$ of G' is the Lie algebra $\mathfrak{g}$ of G considered as a Lie algebra over $\hat{\mathbb{Q}}$. The exponential maps for G and G' are the same. A subgroup H of G' is a Lie subgroup over k if and only it is a Lie subgroup over $\hat{\mathbb{Q}}$ and the $\hat{\mathbb{Q}}$-Lie subalgebra corresponding to it is a k-subspace of $\mathfrak{g}(= \mathfrak{g}')$.*

The lemma is a restatement of 4.31 (4) combined with 5.2 (4). One consequence of the lemma is:

5.17. Proposition. *Let G, H be Lie groups over k. A $\hat{\mathbb{Q}}$ analytic homomorphism $f : G \to H$ of Lie groups over k is analytic over k, if and only if the induced map $\dot{f} : \mathfrak{g} \to \mathfrak{h}$ is linear over k.*

5.18. Lemma. *Let G, H be Lie groups, $F : G \to H$ an analytic group homomorphism and $(B, j : B \to H)$ a Lie subgroup of H. Then there is a Lie subgroup $(A, i : A \to G)$ of G such that $i(A) = F^{-1}(j(B))$.*

$(A, j : A \to G)$ is the *inverse image* of $(B, i : B \to H)$ under F.

5.19. Proof. Let $A = \{(g, h) \in G \times B|\ F(g) = j(h)\}$. Then A is a Lie subgroup of $G \times B$ over $\hat{\mathbb{Q}}$ (Cartan's theorem); its Lie algebra (over $\hat{\mathbb{Q}}$), $\{(X, Y) \in \mathfrak{g} \times \mathfrak{b}|\ dF_1(X) = dj_1(Y)\}$ is a k-linear subspace of $\mathfrak{g} \times \mathfrak{b}$. Hence A is a Lie group over k. If i denotes the restriction of the Cartesian projection $p_G : G \times B \to G$ to A, $(A, i : A \to G)$ is a Lie subgroup of G (over k) with the requisite property.

5.20. We continue with the notation introduced in Chapter 4 (in particular in 4.34–4.36). In that chapter, we associated to a Lie group G a Lie algebra $\mathfrak{g}$. If $(H, i : H \to G)$ is a Lie subgroup and $\mathfrak{h}$ its Lie algebra then the differential di_1 of i at 1 is an injective Lie algebra morphism of $\mathfrak{h}$ in $\mathfrak{g}$. We identify the image of $\mathfrak{h}$ which is a Lie subalgebra of $\mathfrak{g}$ with $\mathfrak{h}$: $\mathfrak{h}$ denotes the Lie algebra of H as well as its image in $\mathfrak{g}$ under di. This assignment of a Lie subalgebra of G to each Lie subgroup $(H, i : H \to G)$ has a converse which is the main result of this chapter. But before we proceed to formulating and proving the converse, we establish some interesting facts about Lie subgroups of Lie groups.

5.21. Theorem. *Assume that k is non-archimedean. Let G be a Lie group and $(H, i : H \to G)$ be a Lie subgroup of G. Let G' be an open subgroup of G of exponential type and $(G', \exp_{G'}^{-1} : G' \to B_d(0))$ be a good chart on G' (cf. 4.42). Let $\mathfrak{h}$ be the Lie subalgebra of $\mathfrak{g}$ corresponding to H. Then for $0 < d' \le d$, $H' = \exp_G(\mathfrak{h} \cap B_{d'}(0))$ is a compact Lie subgroup of G' of exponential type with $\mathfrak{h}$ as the corresponding Lie subalgebra of $\mathfrak{g}$. Further $(H', \exp_{G'}^{-1}|_{H'} : H' \to \mathfrak{h} \cap B_{d'}(0))$ is a good chart on H'. Moreover $i^{-1}(H')$ is an open subgroup of H.*

5.22. Proof. We have identified $\mathfrak{g}$ with k^n through a basis of $\mathfrak{g}$. Let $\mathfrak{o}_{\mathfrak{h}} = \mathfrak{o}_k^n \cap \mathfrak{h}$ and $\mathfrak{v}$ a $\mathfrak{o}_k$-submodule of $\mathfrak{o}_k^n$ such that $\mathfrak{o}_k^n = \mathfrak{o}_{\mathfrak{h}} \oplus \mathfrak{v}$ (cf.1.94). Let V denote the k-linear span of $\mathfrak{v}$; then $\mathfrak{g} = \mathfrak{h} \oplus V$. Moreover one has for $h \in \mathfrak{h}$ and $v \in V$, $||h + v|| = \max(||h||, ||v||)$. Consider now the map $(h, v) \rightsquigarrow \exp_G(h) \cdot \exp_G(v)$ of $(B_d(0) \cap \mathfrak{h}) \times (B_d(0) \cap V)$ (denoted $\mathcal{E}$). Then for some $d_1' > 0$, $d_1' \le d$, $\mathcal{E}|_{(B_{d_1'}(0) \cap \mathfrak{h}) \times (B_{d_1'}(0) \cap V)}$ is an analytic diffeomorphism onto an open subset G_1' of G' and $i^{-1}(\mathcal{E}((B_{d_1'}(0) \cap \mathfrak{h}) \times \{0\}))$ is an open subgroup of H in its Lie group topology. Now $\mathcal{E}' = \exp_G^{-1} \circ \mathcal{E}$ is the restriction of $F (= \exp_G^{-1} \circ \mu \circ (\{\exp_G \times \exp_G) : B_d(0) \times B_d(0) \to B_d(0))$. As $(G', \exp_G^{-1} : G' \to B_d(0))$ is a good chart on G', the Taylor series at $(0,0)$ of F converges uniformly to F on $B_d(0) \times B_d(0)$, so does that of $\mathcal{E}'|_{(\mathfrak{h} \cap B_d(0)) \times (\mathfrak{h} \cap B_d(0))}$. Let p denote the Cartesian projection of $\mathfrak{g}$ on V following the direct sum decomposition $\mathfrak{g} = \mathfrak{h} \oplus V$. Then one has evidently, $p \circ \mathcal{E}' = 0$ on $(\mathfrak{h} \cap B_{d_1'}(0)) \times (\mathfrak{h} \cap B_{d_1'}(0))$ and hence its Taylor series at $(0,0)$ is identically zero. It follows that $p \circ \mathcal{E}' = 0$ on all of $\mathfrak{h} \cap B_d(0) \times \mathfrak{h} \cap B_d(0)$. Thus for $0, d' \le d$, $H' = \exp_G(\mathfrak{h} \cap B_d'(0))$ is a Lie subgroup of G of exponential type with $\mathfrak{h}$ as its Lie algebra. Moreover $(H', \exp_G^{-1}|_{H'} : H' \to (\mathfrak{h} \cap B_d'(0)))$ is a good chart on H'. Since $\exp_G(\mathfrak{h} \cap B_{d'}(0))$ is open in H and contained in H', the last assertion follows.

The next result we prove (valid over any local field) will be used later in several contexts.

5.23. Theorem. *Let $(H', i' : H' \to G)$ be a Lie subgroup of G, and H a subgroup of G containing $i'(H')$ such that $H/i'(H')$ is countable. Then H carries a natural structure of a Lie group such that $(H, i : H \to G)$ is a Lie subgroup of G where i is the inclusion of H in G. i' is the restriction of i to H'. Moreover H' is an open subgroup in the manifold topology on H. The inclusion of H' in H induces an isomorphism of their Lie algebras.*

5.24. Proof. As H' contains an open subgroup of U of exponential type and H'/U is countable, we may assume that H' is of exponential type. Let $\mathfrak{h}'$ be the Lie subalgebra of $\mathfrak{g}$ corresponding to H'. Let $d' > d > 0$ be such that $\exp_G$ is an analytic diffeomorphism of $B_{d'}(0)$ on an open neighbourhood U of the identity in G and if $V = \exp_G(\mathfrak{h} \cap B_d(0))$, $V \cdot V^{-1} \subset \exp_G(\mathfrak{h} \cap B_{d'}(0))$. Suppose now that $X \in \mathfrak{h}'$ and $g \in H$. Then $g \cdot \exp_G(t \cdot X) \cdot g^{-1} \in H$ for all

$t \in k$ with $|t|_k \leq \delta$ for a suitable δ. Since H' admits a countable dense set, there is a countable set $\Xi \subset H$ such that $i'(\exp_{H'}(\mathfrak{h}' \cap B_d(0)) \cdot \Xi) = H$. Since $\{t \in k| \; |t|_k \leq \delta\}$ is uncountable we see that there exist $u, u' \in \{t \in k| \; |t|_k \leq \delta\}$ with $u \neq u'$ and $\xi \in \Xi$ such that $g \cdot \exp_G(u \cdot X) \cdot g^{-1}$ and $g \cdot \exp_G(u' \cdot X) \cdot g^{-1}$ are both in $i'(\exp_{H'}(\mathfrak{h}' \cap B_d(0))) \cdot \xi$. In other words,

$$g \cdot \exp_G((u' - u) \cdot X) \cdot g^{-1} \; =$$

$$g \cdot \exp_G(u' \cdot X) \cdot g^{-1} \cdot g \cdot \exp_G(-u \cdot X) \cdot g^{-1} \in \exp_G(B_d(0) \cap \mathfrak{h}).$$

Equivalently,

$$\exp_G((u' - u) \cdot Ad(g)((u - u') \cdot X)) = \exp_G(Z)$$

for some $Z \in B'_d(0) \cap \mathfrak{h}'$; and since $\exp_G$ is injective on $B'_d(0)$, $Ad(g)(X) = (u' - u)^{-1} \cdot Z$. Thus $Adg(X) \in \mathfrak{h}'$ for every $X \in \mathfrak{h}'$ and $g \in H$. It follows that H is contained in the normalizer $N(\mathfrak{h}') = \{g \in G| \; Ad(g)(\mathfrak{h}') = \mathfrak{h}'\}$ of $\mathfrak{h}'$. So the proposition follows once it is proved *under the additional hypothesis that* $H \subset N(\mathfrak{h}')$ which we assume from now on.

Let $\Xi \subset H$ be a (countable) set which maps bijectively on to H/H' so that H is the disjoint union $\coprod_{\xi \in \Xi} \xi \cdot H'$. We transport the analytic structure on H' to each of the $\xi \cdot H'$, $\xi \in \Xi$ through the bijection $H' \ni h \rightsquigarrow \xi \cdot h$. We thus get an analytic structure on H. We assert that in this analytic structure the multiplication in H is analytic. Let $a, b \in H$ be any points. Then we need to show that the map $(x, y) \rightsquigarrow a \cdot x \cdot b \cdot y$ is analytic in a neighbourhood $U \times U$ of $(1, 1)$ in $H' \times H'$. We have $a \cdot x \cdot b \cdot y = a \cdot b \cdot (b^{-1} \cdot x \cdot b) \cdot y$. Now if U is a sufficiently small neighbourhood of 1 in H', U as well as $b^{-1} \cdot U \cdot b \subset H'$ are open subsets in H' and inherit their manifold topology from G. This follows from the fact that for fixed $h \in H$, $Ad_G(h)(\mathfrak{h}) \subset \mathfrak{h}$ for and $h \cdot \exp_G(X) \cdot h^{-1} = \exp_G(Ad_G(h)(X))$ for all $X \in \mathfrak{h}$ sufficiently close to 0 in $\mathfrak{h}$. It follows that the map $x \rightsquigarrow (b^{-1} \cdot x \cdot b)$ is an analytic map of U into H' (see 3.29). To prove that the multiplication in H is analytic, we argue as follows. Let x_n, y_n be sequences in H tending to ξ, η respectively in H for the manifold topology introduced above. Then one has $x_n \in \xi \cdot H'$ and $y_n \in \eta \cdot H'$ so that $x_n = u_n \cdot \xi$ and $y_n \cdot \eta$ with $u_n, v_n \in H'$ tending to 1 in the manifold topology of H'. We then have $x_n \cdot y_n = u_n \cdot \xi \cdot v_n \cdot \eta = R_{\eta^{-1}}(u_n \cdot \text{Int}\xi(v_n))$. As $\text{Int}(\xi)$ is continuous as also $R_{\eta^{-1}}$ and the multiplication in H' is continuous (in its manifold topology) as well, we see that $x_n \cdot y_n$ tends to $\xi \cdot \eta$, proving that the multiplication in H is continuous in its manifold topology. That the multiplication is analytic now follows from 3.29.

We record as a theorem for future use the following special case of Theorem 5.23; in fact in the course of proving Theorem 5.23 we have proved Theorem 5.25 below.

5.25. Theorem. *Let $\mathfrak{h}$ be a Lie subalgebra of $\mathfrak{g}$. Let Γ be any countable subgroup of $N(\mathfrak{h}) = \{g \in G|\ Ad_{\mathfrak{g}}(g)(\mathfrak{h}) = \mathfrak{h}\}$. Let $(H, i : H \to G)$ be a Lie subgroup of G with $\mathfrak{h}$ as its Lie algebra. Then the group H^* generated by Γ and $i(H)$ has a natural structure of a Lie group making it into a Lie subgroup of G with $\mathfrak{h}$ as its Lie algebra. H is an open subgroup of H^* in its Lie group topology. Conversely, every Lie subgroup of G with $\mathfrak{h}$ as its Lie algebra is of the form H^* for a suitable Lie subgroup H of exponential type and a countable subgroup Γ of $N(\mathfrak{h})$.*

We now state and prove the Fundamental Theorem of Lie Theory.

5.26. Fundamental Theorem of Lie Theory. *Let G be a Lie group and $\mathfrak{h}$ a Lie subalgebra of $\mathfrak{g}$. Then there is a Lie subgroup $(H, i : H \to G)$, H of exponential type, with $i(H) \subset G^o$, an open subgroup of G of exponential type, with $\mathfrak{h}$ as the Lie algebra of H. When k is non-archimedean and G is of exponential type with $(G, \exp^{-1} : G \to B_d(0))$ as a good chart, H can be chosen to be $\exp(\mathfrak{h} \cap B_d(0))$; it is compact and the topology induced on H from G is the topology on H as a Lie group. When k is archimedean, H being of exponential type is (necessarily) connected and $i(H)$ is the subgroup of G generated by $\exp_G(\mathfrak{h})$.*

5.27. Proof of 5.26 – First Reduction. Replacing G by an open subgroup, we may assume that G *is of exponential type.* Set

$$N = \{g \in G|\ Ad(g)(\mathfrak{h}) = \mathfrak{h}\}.$$

Then by Cartan's theorem (5.3), N is a Lie subgroup over $\hat{\mathbb{Q}}$. The Lie subalgebra of $\mathfrak{n}$ of $\mathfrak{g}$ corresponding to N is

$$\{X \in \mathfrak{g}|\ [X, \mathfrak{h}] \subset \mathfrak{h}\}.$$

Evidently $\mathfrak{n}$ is a Lie subalgebra of $\mathfrak{g}$ over k. Thus N is a Lie subgroup over k. Now $\mathfrak{h}$ is evidently contained in $\mathfrak{n}$ and is an ideal in it. Thus we may assume – *replacing G by an open subgroup of N of exponential type – that $\mathfrak{h}$ is an ideal in $\mathfrak{g}$ which is $Ad(G)$-stable.*

5.28. Proof of 5.26 – Second Reduction. Since $\mathfrak{h}$ is $Ad(G)$-stable, we obtain a representation ρ of G on $\mathfrak{g}/\mathfrak{h}$. Then the kernel B of ρ, being a closed subgroup of G is a Lie subgroup over $\hat{\mathbb{Q}}$. Its Lie algebra is easily seen to be $\mathfrak{b} = \{X \in \mathfrak{g}|\ [X, Y] \in \mathfrak{h},\ \forall\ Y \in \mathfrak{g}\}$ and is thus a k-subalgbra of $\mathfrak{g}$. It follows that B is a Lie subgroup of G over k. Then one has $[\mathfrak{b}, \mathfrak{b}] \subset \mathfrak{h}$. Evidently $\mathfrak{b}$ contains $\mathfrak{h}$. We may hence replace G by B. It follows that we are reduced to proving the theorem in the special case when $\mathfrak{h}$ *is an ideal in $\mathfrak{g}$ containing* $[\mathfrak{g}, \mathfrak{g}]$.

5.29. Proof of 5.26 Under the Assumption $\mathfrak{h} \supset [\mathfrak{g}, \mathfrak{g}]$. Fix a basis of $\mathfrak{g}/\mathfrak{h}$ so that we may identify the natural map $\mathfrak{g} \to \mathfrak{g}/\mathfrak{h}$ with a q-tuple of linear forms $\{f_r : \mathfrak{g} \to k |\ 1 \leq r \leq q\}$ (where $q = \dim_k \mathfrak{g}/\mathfrak{h}$) on $\mathfrak{g}$. The f_i considered as linear forms on $\mathfrak{g}$ (vanishing on $[\mathfrak{g}, \mathfrak{g}]$) define a q-tuple of left and right translation invariant 1-forms $\{\omega_r |\ 1 \leq r \leq q\}$ on G. The fact that the f_r vanish on $[\mathfrak{g}, \mathfrak{g}](\subset \mathfrak{h})$ implies that the forms ω_r are closed (cf. 3.63). By the Poincaré lemma (see 3.64), we can find an open covering $\mathfrak{U} = \{U_i |\ i \in I\}$ of G and analytic functions $\{f_{ri} : U \to k |\ i \in I,\ 1 \leq r \leq q\}$ such that $df_{ri} = \omega_r|_{U_i}$. At this point we separate out the archimedean and non-archimedean cases.

5.30. Conclusion of Proof When k is Non-Archimedean. In this case we may assume that one of the U_i which contains 1 is a compact open subgroup of G which we denote U and that $f_r \overset{\text{def}}{=} f_{ri}$ vanish at 1 for $1 \leq r \leq q$. Let $m : U \times U \to U$ (resp. $p_1 : U \times U \to U$, resp. $p_2 : U \times U \to U$) denote the multiplication in U (resp. Cartesian projection on the first factor, resp. Cartesian projection on the second factor). Then for $1 \leq r \leq q$,

$$d(f_r \circ m - f_r \circ p_1 - f_r \circ p_2) = 0$$

which implies that the function $(x, y) \rightsquigarrow f_r(x \cdot y) - f_r(x) - f_r(y)$ on $U \times U$ is locally constant. Since $f_r(1) = 0$, it follows that when restricted to a suitable compact open subgroup U' of U, $f = \{f_r\}_{1 \leq r \leq q}$ is an analytic group homomorphism of U' in k^r. The kernel H of $f|_{U'}$ is a closed Lie subgroup of G with $\mathfrak{h}$ as its Lie algebra. The more precise assertions in the non-archimedean case follow from Theorem 5.21.

5.31. Conclusion of Proof When k is Archimedean. Consider first the case when G is simply connected. For $i, j \in I$ and $1 \leq r \leq q$ (as in 5.29) $d(f_{ri} - f_{rj}) = 0$ on $U_i \cap U_j$. It follows that $c_{rij} = f_{ri} - f_{rj}$ is locally constant. Since G is simply connected, by Theorem 1.78, $\mathcal{U}$ admits a refinement $(\mathcal{V} = \{V_j|\ j \in J\}, \rho : J \to I)$ such that there are locally constant functions u_{rj} on U_j such that $u_{rj} - u_{rj'} = c_{r\rho(j)\rho(j')}$ for all $j, j' \in J$. It follows that $\Phi_{rj} = f_{r\rho(j)} - u_{rj}$, $j \in J$ patch up to give an analytic function $f_r : G \to k$. We may assume (by adding a constant if need be) that $f_r(1) = 0$. Arguing as in 5.29, we conclude that $f = \{f_r\}_{1 \leq r \leq q} : G \to k^q$ is an analytic group homomorphism whose kernel H is a (closed) Lie subgroup with $\mathfrak{h}$ as its Lie algebra.

For a general G, let $p : \tilde{G} \to G$ be the universal covering of G. Then the differential dp of p at 1 is an isomorphism of the Lie algebra $\tilde{\mathfrak{g}}$ (of $\tilde{G}$) on $\mathfrak{g}$. Let $\tilde{\mathfrak{h}}$ be the subalgebra of $\tilde{\mathfrak{g}}$ which maps to $\mathfrak{h}$ under dp. Now there is a closed Lie subgroup $\tilde{H}$ over k of $\tilde{G}$ such that $(\tilde{H}, \tilde{i} : \tilde{H} \to \tilde{G})$ with $\tilde{i}$ inducing an isomorphism of the Lie algebra of $\tilde{H}$ on $\tilde{\mathfrak{h}}$. Let $H = \tilde{H}/(\tilde{H} \cap (\text{kernel } p))$.

As $\tilde{H} \cap$ (kernel p) is a closed discrete subgroup of $\tilde{H}$, H carries a natural structure of a Lie group (cf. 4.10 (9)); and $p \circ \tilde{i}$ factors through H to give an injective analytic map i of H in G. Hence $(H, i : H \to G)$ is a Lie subgroup with $\mathfrak{h}$ as its Lie algebra. This completes the proof of Theorem 5.26.

If $\mathfrak{h}$ in the Theorem 5.26 is an ideal, we have in view of 5.21.

5.32. Theorem. *If G is of exponential type and $\mathfrak{h}$ is an ideal in $\mathfrak{g}$, a Lie subgroup $(H, i : H \to G)$ corresponding to $\mathfrak{h}$ can be chosen to be normal in G. When k is archimedean, the connected component of the identity in H is normal in G (which is necessarily connected). When k is non-archimedean and $(G, \exp^{-1} : G \to B_d(0))$ is a good coordinate system, H can be taken to be $\exp(\mathfrak{h} \cap B_d(0))$.*

5.33. Remarks. (1) The Lie subgroup (of exponential type with a given subalgebra $\mathfrak{h}$ of $\mathfrak{g}$ as its Lie algebra is not unique when k is non-archimdean: we have seen that if H is one such Lie subgroup, there is a fundamental system of open neighbourhoods of 1 in H which are all open compact subgroups of H of exponential type and are thus all Lie subgroups with $\mathfrak{h}$ as the Lie algebra.

(**2**) In the archimedean case, a Lie group of exponential type is necessarily connected, which ensures the uniqueness of the Lie subgroup (of exponential type) corresponding to a Lie subalgebra. In the sequel, we will speak of *the* Lie subgroup corresponding to a Lie subalgebra in the case of archimedean k and that means the Lie subgroup is connected.

(**3**) If H is a Lie subgroup of G corresponding to a Lie subalgebra $\mathfrak{h}$, then $H \subset N(\mathfrak{h}) = \{g \in G | \ Ad(g)(\mathfrak{h}) = \mathfrak{h}\}$.

5.34. Corollary to Theorem 5.26. *Let G and H be Lie groups and $\mathfrak{g}$ and $\mathfrak{h}$ their respective Lie algebras. If k is archimedean, assume that G is connected and simply connected. Then given any Lie algebra morphism $f : \mathfrak{g} \to \mathfrak{h}$, there is an open subgroup G' of G and a homomorphism $\tilde{f} : G' \to H$ such that f is the map induced by $\tilde{f}$.*

5.35. Proof. Let $\mathfrak{g}_f = \{(X, f(X))| \ X \in \mathfrak{g}\}$ ($\subset \mathfrak{g} \times \mathfrak{h}$). Then $\mathfrak{g}_f$ is evidently a Lie subalgebra of $\mathfrak{g} \times \mathfrak{h}$, the Lie algebra of $G \times H$. By Theorem 5.26, there is a lie subgroup $(G_f, i_f : G_f \to G \times H)$ of exponential type) such that $\dot{i}_f$ is an isomorphism of the Lie algebra of G_f on $\mathfrak{g}_f$. Let $\pi_G : G \times H \to G$ be the Cartesian projection. Then $\Phi = \pi_G \circ i_f$ is a homomorphism of G_f to G which induces an isomorphism of their Lie algebras. It follows that Φ is a covering projection. When k is archimedean, we have assumed that G is simply connected; this implies that Φ is an isomorphism and we see that $p_H \circ i_f \circ \Phi^{-1}$ is a homomorphism of G in H and may be taken as $\tilde{f}$. Here p_H

is the Cartesian projection of $G \times H$ on H. When k is non-archimedean, the homomorphism Φ restricted to a suitable compact open subgroup G_{f1} of G_f is an isomorphism on to a compact open subgroup G_1 of G. If Ψ is the inverse of $\Phi|_{G_{f1}} : G_{f1} \to G_1$, $\tilde{f} = \pi_H \circ i_f \circ \Psi$ satisfies the requirements of the corollary.

5.36. Remark. When k is archimedean, the condition that G is simply connected is essential. The Lie algebra of the real Lie group $\mathbb{S}^1$, the unit circle in $\mathbb{C}$, has a natural identification with $\mathbb{R}$ such the exponential map $\exp_{\mathbb{S}^1} : \mathbb{R} \to \mathbb{S}^1$ is the map $\mathbb{R} \ni t \rightsquigarrow e^{2\pi \cdot i \cdot t}$. It is easy to see that if α is irrational, the map $x \rightsquigarrow \alpha \cdot x$ of $\mathbb{R}$ to $\mathbb{R}$ is not induced by an analytic homomorphism of $\mathbb{S}^1$ into itself.

5.37. When k is non-archimedean, the proof of Theorem 5.30 given above actually gives the following more precise statement: Let $(G, \exp_G^{-1} : G \to B_d(0; \mathfrak{g}))$ and $(H, \exp_H^{-1} : H \to B_d(0; \mathfrak{h}))$ be good coordinate charts on G and H respectively; then $B_o = \exp_{G \times H}(B_d(0, \mathfrak{b}))$ is a compact Lie subgroup of $G \times H$ such that $p_G|_{B_o}$ is an analytic diffeomorphism of B_o on G and $\tilde{f} = p_H \circ i \circ (p_G|_B)^{-1}$ is an analytic homomorphism of G into H which induces f.

5.38. Automorphism Group of a Lie Group (k Archimedean). In 5.10 we saw that for any Lie algebra $\mathfrak{g}$ over k, the automorphism group $A(\mathfrak{g})$ has a natural structure of a Lie group over k making it a Lie subgroup over k of $GL_k(\mathfrak{g})$. We saw also that its Lie algebra $\mathfrak{a}(\mathfrak{g})$ is the Lie algebra of all derivations of $\mathfrak{g}$ over k. Now let G be a Lie group over k with $\mathfrak{g}$ as its Lie algebra. We saw (5.12–5.13) that if k is non-archimedean, $A(G)$, the automorphsim group of G, has a natural identification with a compact open subgroup of $A(\mathfrak{g})$ and is thus a Lie subgroup over k of $GL_k(\mathfrak{g})$.

Assume now that k is archimedean and G is a *connected* Lie group (over k) with $\mathfrak{g}$ as its Lie algebra. Let $\tilde{G}$ be a universal covering group of G and $p : \tilde{G} \to G$ the covering map. Let Γ be the kernel of p. Then Γ is a closed discrete subgroup of $\tilde{G}$. From 5.34 we see that if $g \in A(\mathfrak{g})$, there is an automorphism $\tilde{g} \in A(\tilde{G})$ which induces g on $\mathfrak{g}$. Thus the map $A(\tilde{G}) \to A(\mathfrak{g})$ is surjective. On the other hand we saw that $A(\tilde{G}) \to A(\mathfrak{g})$ is injective (5.12–5.13); it follows that $A(\tilde{G}) \simeq A(\mathfrak{g})$ and is thus a Lie group over k. Moreover, it is easily seen (using 5.10) that the map $(a, g) \rightsquigarrow a(g)$ is an analytic action (over k) of $A(\tilde{G})$ on $\tilde{G}$. Let $S(G)$ be the semi-direct product $A(\mathfrak{g})$ and $\tilde{G}$ over k (cf. 4.10 (11)). The Lie algebra $\mathfrak{s}(G)$ of $S(G)$ is evidently $\mathfrak{a}(\mathfrak{g}) \propto \mathfrak{g}$. It is easy to see (using Lemma 5.3) that the Lie algebra $\mathfrak{a}(G)$ of $A(G)$ is $\{X \in \mathfrak{a}(\mathfrak{g})| \ Ad_{S(G)}(c)(X) = X, \ \forall \ c \in \Gamma\}$ which is evidently a k-subspace of $\mathfrak{a}(\mathfrak{g})$. It follows that $A(G)$ is a Lie subgroup over k (Lemma 5.16).

5.39. Lie Subgroups of $GL(V)$. Let k be a non-archimedean local field with residue field characteristic p. Let V be a finite dimensional vector space over k. In this case we can prove a more refined version of 5.26, but before stating the result we make a few observations about the exponential map $\exp_{GL(V)}$ in the case of $GL(V)$. There is a finite extension K of k such that all the eigen-values of all the elements of $GL(V)$ lie in K (cf. 1.95). Recall that the spectral radius $\sigma(X)$ of $X \in \mathfrak{gl}(V)$ is

$$\max\{|\lambda|_K| \ \lambda \ an \ eigen\text{-}value \ of \ X\}.$$

It equals $\lim_{n\to\infty}\|X^n\|_k^{1/n}$ where $\|\cdot\|_k$ is any norm on $\mathfrak{gl}(V)$ (obtained from a norm on the k-vector space V) – note that if $\|\cdot\|$ and $\|\cdot\|'$ are two such norms, there is a constant $C > 1$ such that

$$C^{-1} \cdot \|Y\| \le \|Y\|' \le C \cdot \|Y\|$$

for all $Y \in \mathfrak{gl}(V)$. Let

$$\Omega = \Omega_{\mathfrak{gl}(V)} = \{X \in \mathfrak{gl}(V)| \ \sigma(X) \le p^{-2/(p-1)}\}.$$

As the eigen-values of $T \in \mathfrak{gl}(V)$ vary continuously with T, we see that Ω is open in $\mathfrak{gl}(V)$. The exponential series $\sum_{n\in\mathbb{N}} X^n/n!$ is easily seen to converge absolutely and uniformly on compact subsets of Ω (compare 4.34): the series is indeed the Taylor series of the exponential function at 0. Now, let $\Omega^* = \{1+X| \ X \in \mathfrak{gl}(V), \sigma(X) \le p^{-2/(p-1)}\}$. Then Ω^* is open subset of $GL(V)$ (and hence of $\mathfrak{gl}(V)$ as well. The series $\ln(A) = \sum_{n=1}^{\infty}(-1)^{(n+1)}(A-1)^n$ converges absolutely and uniformly on compact subsets of Ω^* to an analytic function $\ln : \Omega^* \to \Omega$. It is not difficult to see that $\exp(\Omega) \subset \Omega^*$ and $\ln(\Omega^*) \subset \Omega$; further $ln\circ\exp = 1_\Omega$ and $\exp\circ\ln = 1_{\Omega^*}$ (cf. 2.12). Thus exp is an analytic diffeomorphism of Ω on Ω^*.

5.40. Theorem. *Assume that k is non-archimedean with residue field characteristic p. Let $\mathfrak{h}$ be a Lie subalgebra of $\mathfrak{gl}(V)$ and $\Omega_\mathfrak{h} = \Omega_{\mathfrak{gl}(V)} \cap \mathfrak{h}$. Let H be the subgroup of $GL(V)$ generated by $\exp_{\mathfrak{gl}(V)}(\Omega_\mathfrak{h})$. Then H has a structure of a Lie group making its inclusion in $GL(V)$ a Lie subgroup of $GL(V)$ corresponding to the Lie subalgebra $\mathfrak{h}$.*

The subgroup H as in the theorem is the *canonical Lie subgroup of $GL(V)$ corresponding to* $\mathfrak{h}$.

5.41. Proof. Let $W = \{(X,Y) \in \Omega_\mathfrak{h} \times \Omega_\mathfrak{h}| \ \exp_{GL(V)}(X) \cdot \exp_{GL(V)}(Y) \in \Omega^*_{GL(V)}\}$. Then W is an open subset of $\mathfrak{h} \times \mathfrak{h}$. Let $F = \ln\circ m\circ(\exp_{GL(V)} \times \exp_{GL(V)})|_W$. Let $\pi : \mathfrak{gl}(V) \to \mathfrak{gl}(V)/\mathfrak{h}$ be the natural map. Then the Taylor series of the analytic function $\pi\circ F$ at $(0,0)$ converges to the function $\pi\circ F$ on W. Now if Φ is a suitably small nighbourhood of 0 in $\mathfrak{h}$, $U = \exp_{GL(V)}(\Phi)$ is a compact Lie subgroup of $GL(V)$ with $\mathfrak{h}$ as its Lie algebra and $\ln(U) \subset \mathfrak{h}$. It follows that $\pi\circ F(\Phi \times \Phi) = \{0\}$ and hence the Taylor series at $(0,0)$ is identically zero. This implies that $\pi\circ F$ is identically zero; in other words $\ln(F(W)) \subset \mathfrak{h}$.

Suppose now that Δ is a countable dense set of $\Omega_{\mathfrak{h}}^* = \exp_{GL(V)}(\Omega_{\mathfrak{h}})$ and $\tilde{\Delta}$ the (countable) group generated by Δ. Let $\{\Omega_{\mathfrak{h},n} \subset \Omega_{\mathfrak{h}}| \ n \in \mathbb{N}\}$ be a fundamental system of neighbourhoods of 0 in $\mathfrak{h}$ such that $\Omega_{\mathfrak{h},n+1} \subset \Omega_{\mathfrak{h},n}$ for all $n \in \mathbb{N}$ and $H_n = \exp_{gl(V)}(\Omega_{\mathfrak{h},n})$ are all compact Lie subgroups of $GL(V)$ with $\mathfrak{h}$ as their Lie algebra. We will now show that $H = H_0 \cdot \tilde{\Delta}$. This will prove the theorem in the light of 5.23. We will prove by induction on n that for any n-tuple $\{A_i \in \Omega_{\mathfrak{h}}^*| \ 1 \leq i \leq n\}$, $A_1 \cdot A_2 \cdots A_n \in H_0 \cdot \tilde{\Delta}$. Suppose that this holds for some n. Let $\{A_i \in \Omega_{\mathfrak{h}}^*| \ 1 \leq i \leq n+1\}$ be an $(n+1)$-tuple. Let $B = A_1 \cdot A_2 \cdots A_n$. Then by induction hypothesis, $B = h \cdot \gamma$ with $h \in H_0$ and $\gamma \in \tilde{\Delta}$. Now, as W is open in $\mathfrak{h} \times \mathfrak{h}$ there is an integer $N \geq 0$ such that $(A_{n+1}, X) \in W$ for all $X \in \Omega_{\mathfrak{h},m}$ for all $m \geq N$. It follows that there exist for all $m \geq N$, $u_m \in H_m$ such that $u_m^{-1} \cdot A_{n+1} = \theta_m \in \Delta$. We then have $B \cdot A_{n+1} = h \cdot g \cdot u_m \cdot \theta$. Now for $Y \in \Omega_{\mathfrak{h}}$ and $Z \in \mathfrak{gl}(V)$, $(Ad_{\mathfrak{gl}(V)}(\exp(Y)))(Z) = Ad(\exp_{\text{End}(\mathfrak{gl}(V))}(ad(Y)))(Z)$ so that the group $\tilde{\Delta}$ normalizes $\mathfrak{h}$. Now choose m so large that $h' = \gamma \cdot u_m \gamma^{-1} \in H_0$. It follows than $B \cdot A_{n+1} = h \cdot h' \cdot \gamma \cdot \theta_m \in H_0 \cdot \tilde{\Delta}$. The argument takes care of the start of the induction at $n = 1$ as well. Hence the theorem.

5.42. Remarks. (1) If the Lie subalgebra $\mathfrak{h}$ consists entirely of nilpotent matrices, $\mathfrak{h} \subset \Omega_{\mathfrak{h}}$ and $W \subset (\mathfrak{h} \times \mathfrak{h})$ so that $H = \exp_{GL(V)}(\mathfrak{h})$.

(2) If we identify k with the Lie subalgebra of scalar matrices, $\Omega \cap k = \{a \in k| \ |a| \leq p^{-2/(p-1)}\}$ and the group H of the above theorem is the group $\{x \in k^\times| \ |x - 1|_k < p^{1/(p-1)}\}$. On the other hand $k^\times$ is also a Lie subgroup corresponding to k.

5.43. Admissible Representations. A representation ρ of a Lie group G over the (local) field k is *admissible* if $\rho(G)$ is contained in the subgroup of $GL(V_\rho)$ generated by $\exp_{GL(V)}(\Omega_{\dot\rho(\mathfrak{g})})$. A Lie subgroup $(G, i : G \to GL(V))$ of $GL(V)$ is *admissible* if i is an admissible representation. Note that every representation of a connected Lie group (over an archimedean field k) is admissible and hence every connected Lie subgroup of $GL(V)$ is admissible.

It is clear from the definition that if $\mathfrak{h}$ is a Lie subalgebra of $GL(V)$, there is an admissible Lie subgroup $(H, i : H \to GL(V))$ with $\mathfrak{h}$ as its Lie algebra.

Note that if σ and τ are admissible representations of a Lie group G, then $\sigma \otimes \tau$ is admissible. If ρ is an admissible representation, any sub-representation or quotient representation of ρ is admissible. It follows that the tensor, symmetric and exterior powers, of an admissible representation are admissible.

5.44. Proposition. *Let ρ be an admissible representation of the Lie group G. Then a vector subspace of V_ρ is stable under G if and only if it is stable under $\mathfrak{g}$.*

5.45. The Lie algebra of a Lie group carries a lot of information about the Lie group. Some of the results in the sequel will illustrate this.

5.46. Theorem. *Let $\mathfrak{c}$ be the centre of $\mathfrak{g}$. Then there is an open subgroup G_o of G such that $\mathfrak{c}$ is the Lie algebra of the centre C_o of G_o. When k is archimedean, G_o may be taken as the connected component of the identity in G. When k is non-archimedean G_o may be taken as any open subgroup of exponential type. If Ad_G is admissible, G_o may be taken as G.*

This follows from the definitions and 5.21. We next prove the following.

5.47. Theorem. *Let G be a Lie group and $(H_i, \alpha_i : H_i \to G)$, $i = 1, 2$ be Lie subgroups. Let $\mathfrak{h}_i$ be the subalgebra of $\mathfrak{g}$ corresponding to H_i. Let $\mathfrak{h} = \mathfrak{h}_1 \cap \mathfrak{h}_2$. Then there is a Lie subgroup $(H, \alpha : H \to G)$ such that $\mathfrak{h}$ is the corresponding Lie subalgebra of $\mathfrak{g}$ and $\alpha(H)$ is the subgroup $\alpha_1(H_1) \cap \alpha_2(H_2)$.*

5.48. Proof. Let $H = \{(x_1, x_2) \in H_1 \times H_2 | \alpha_1(x_1) = a_2(x_2)\}$. Then H is a closed subgroup of the Lie group $H_1 \times H_2$ and hence $(H, j : H \subset H_1 \times H_2)$ is a Lie subgroup over $\hat{\mathbb{Q}}$. It is easy to see that the Lie subalgebra $\mathfrak{h}$ of $H_1 \times H_2$ corresponding to $(H, j : H \subset H_1 \times H_2)$ is a k-linear subspace. Thus $(H, j : H \subset H_1 \times H_2)$ is a Lie subgroup over k. Let $\alpha = \alpha_1 \circ p_{H_1}|_H = \alpha_2 \circ p_{H_2}|_H$ where for $i = 1, 2$, p_{H_i} is the Cartesian projection of $H_1 \times H_2$ on H_i. Then $(H, \alpha : H \to G)$ is Lie subgroup with the requisite properties.

5.49. Theorem. Let G be a Lie group and $(G_i, \alpha_i : G_i \to G)$, $i = 1, 2$ be Lie subgroups. Let $\mathfrak{g}_i$ be the Lie algebra of G_i and $\mathfrak{g}'_i = \dot{\alpha}_i(\mathfrak{g}_i)$ the subalgebra of $\mathfrak{g}$ corresponding to G_i. Assume that $\alpha_1(G_1)$ normalizes $\alpha_2(G_2)$. Then there is a Lie subgroup $(H, \alpha : H \to G)$ with $\mathfrak{h} \simeq \mathfrak{h}' = \mathfrak{g}'_1 + \mathfrak{g}'_2$ as the corresponding Lie subalgebra of $\mathfrak{g}$ and $\alpha(H) = \alpha_1(G_1) \cdot \alpha_2(G_2)$.

5.50. Proof of Theorem 5.49: Step 1. Let Ω be an exponential neighbourhood of 0 in $\mathfrak{g}$ chosen such that $\Omega_i = \dot{\alpha}_i^{-1}(\Omega)$ is an exponential neighbourhood of 0 in $\mathfrak{g}_i$. Then for any $X \in \mathfrak{g}'_2$ and $g \in G_1$, there is a $\delta > 0$ such that $\exp_G(Ad_G(\alpha_1(g))(t \cdot X)) \in \alpha_2(G_2)(\overset{\text{def}}{=} G'_2)$ for all $t \in I_\delta(0)$. It follows now from 5.3 that $Ad_G(\alpha_1(g))(\mathfrak{g}') \subset \mathfrak{g}'$ for all $g \in G_1$. This implies that $[\mathfrak{g}'_1, \mathfrak{g}'_2] \subset \mathfrak{g}'_2$. Consequently, $\mathfrak{h}' = \mathfrak{g}'_1 + \mathfrak{g}'_2$ is a Lie subalgebra of $\mathfrak{g}$.

5.51. Proof of Theorem 5.49: Step 2. Let $(H^*, \alpha^* : H^* \to G)$ be a Lie subgroup such that $\dot{i}^*$ maps the Lie algebra $\mathfrak{h}^*$ of H^* isomorphically on $\mathfrak{h}'$. Now there is an open neighbourhood Δ of 0 in $\mathfrak{h}^*$ such that $\Delta \subset \dot{i}^{*-1}(\Omega)$. Let H_o be the (open) subgroup of H^* generated by $\exp_{H^*}(\Delta)$; then $(H_o, \alpha_o = \alpha^*|_{H_o})$ is a Lie subgroup of G whose Lie algebra is naturally isomorphic to $\mathfrak{h}^*$ and $\alpha_o(Ho)$ is contained in the subgroup $\alpha_1(G_1) \cdot \alpha_2(G_2)$. Let Φ (resp. Ψ) be a countable dense subgroup of G_1 (resp. G_2). Let Θ be the countable subgroup of $\alpha_2(G_2)$ generated by $\{x \cdot \alpha_2(\Psi) \cdot x^{-1} | x \in \alpha_1(\Phi)\}$ and $\Gamma = \alpha_1(\Phi) \cdot \Theta$. Then $\alpha_o(H_o) \cdot \Gamma = \alpha_1(G_1) \cdot \alpha_2(G_2)$, as is easily seen. It now follows from

Theorem 5.25 that $a_1(G_1) \cdot \alpha_2(G_2)$ carries a natural structure of a Lie group H in which H_o is an open subgroup, the inclusion $\alpha : H \to G$ is analytic and $\alpha(H) = \alpha_1(G_1) \cdot \alpha_2(G_2)$. Hence the theorem.

5.52. Some Definitions and Notation. In the sequel, we set $i + 1 = 1$ if $i = 2$. For i=1,2 let $(G_i, \alpha_i : G_i \to G)$ be Lie subgroups of G such that $\alpha_i(G_i) = G'_i$ are normal subgroups of G. Let $\dot{\alpha}_i : \mathfrak{g}_i \to \mathfrak{g}$ be the Lie algebra morphism induced by α_i. Then as is easily seen, $\mathfrak{g}'_i = \dot{\alpha}_i(\mathfrak{g}_i)$ are ideals in $\mathfrak{g}$ and hence so is $\mathfrak{h}' = [\mathfrak{g}'_1, \mathfrak{g}'_2]$. For $i = 1, 2$, let $\mathfrak{h}'_i$ be the k-linear span of $\{Ad(g)(X) - X \mid X \in \mathfrak{g}'_i, g \in G_{i+1}\}$ and $\mathfrak{h}^* = \mathfrak{h}'_1 + \mathfrak{h}'_2$. Then in view of the fact that G_i are normal in G, we see that the $\mathfrak{h}'_i$, $i = 1, 2$, and $\mathfrak{h}^*$ are stable under $Ad(G)$ and are hence ideals in $\mathfrak{g}$. We note that if both the G_i, $i = 1, 2$ are of exponential type $\mathfrak{h}^* = \mathfrak{h}'$.

Let $c : G \times G \to G$ be the commutator map: for $x, y \in G$,

$$c(x, y) = x \cdot y \cdot x^{-1} \cdot y^{-1}.$$

$c(x, y)$ is also denoted $[x, y]$. For fixed $h \in G$, $c_h : G \to G$ is defined by setting for $g \in G$, $c_h(g) = c(h, g)$. Let H^* be the subgroup generated by $c(G'_1 \times G'_2)$. Note that it is the same as that generated by $c(G'_2 \times G'_1)$.

With these definitions and notation we have the following theorem.

5.53. Theorem. *There is a Lie subgroup $(H, \alpha : H \to G)$ such that $\alpha(H) = H^*$ and $\dot{\alpha} : \mathfrak{h} \to \mathfrak{g}$ is an isomorphism of $\mathfrak{h}$ on the subalgebra $\mathfrak{h}^*$.*

5.54. Proof of Theorem 5.53. Let G_o be an open subgroup of G of exponential type and $d > 0$ be such that $\{\exp_G^{-1} : G_o \to B_d(0)$ is a good coordinate system. Let $(H_o, i_o : H_o \to G)$ be a Lie subgroup of G corresponding to the Lie subalgebra $\mathfrak{h}^*$ so that $\dot{i}_o \mathfrak{h}_o \simeq \mathfrak{h}^*$. We assume d so chosen that $\exp_{H_o}|_{i_o^{-1}(\mathfrak{h}^* \cap B_d(0))}$ is an analytic diffeomorphism onto an open subset U_o of H_o so that $(U_o, \exp_{H_o}^{-1} U_o \to \Omega(\subset \mathfrak{h}^*))$ is a good coordinate chart. Let W be a vector subspace of $\mathfrak{g}$ such that $\mathfrak{g} = \mathfrak{h}^* \oplus W$ and π the projection of $\mathfrak{g}$ on W following this decomposition. Let $g \in G_{i+1}$ and K a compact neighbourhood of 1 in G_{i+1}. Let $u \in g \cdot K$ and consider the map $c_u : G_i \to G$. A simple calculation shows that the tangent map of c_u at $x_i \in G$ takes the tangent vector X_i to G_i at x_i (identified with an element of $\mathfrak{g}'_i$) to

$$Ad(x_i \cdot u \cdot x_i^{-1})(Ad(x_i)(X_i)) - Ad(x_i)(X_i) \in \mathfrak{h}'_{i+1}. \qquad (*)$$

For $i = 1, 2$, let $\{u_{i,j} \mid 1 \leq j \leq r_i\}$ be a finite subset of G_i such that

$$\{Ad_G(u_{i,j})(X) \mid X \in \mathfrak{h}_{i+1}, 1 \leq j \leq r_i\}$$

spans $\mathfrak{h}'_{i+1}$ as a vector space. Let

$$F : \mathcal{G} \overset{\text{def}}{=} (\prod_{1 \leq j \leq r_2} G_1 \times \prod_{1 \leq j \leq r_1} G_2) \to G$$

be the map defined as follows: for $\underline{x}_1 = \{x_{1j} | \ 1 \leq j \leq r_2\} \in \prod_{1 \leq j \leq r_2} G_1$ and
$\underline{x}_2 = \{x_{2j} | \ 1 \leq j \leq r_1\} \in \prod_{1 \leq j \leq r_1} G_2$,

$$F(\underline{x}_1, \underline{x}_2) = (\prod_{1 \leq j \leq r_2} c_{u_{2j}}(x_{1j})) \cdot (\prod_{1 \leq j \leq r_1} c_{u_1j}(x_{2j})).$$

Then from $(*)$ one sees that the differential of F at any point p of $\mathcal{G}$ maps the tangent space at p into the subspace $\mathfrak{h}^*$ of the tangent space $\mathfrak{g}$ to G at the image point. Further our choice of the $\{u_{i,j}\}$ ensures that the tangent map at the identity of $\mathcal{G}$ is onto $\mathfrak{h}^*$; and hence the tangent map is onto $\mathfrak{h}^*$ at all points sufficiently close to the identity in $\mathcal{G}$.

5.55. Proof of Theorem 5.53 (Continued). Now Let V be a subspace of $\mathfrak{g}$ such that $\mathfrak{g}$ is the direct sum of $\mathfrak{h}^*$ and V. Let $\pi : \mathfrak{g} \to V$ be the Cartesian projection of $\mathfrak{g}$ on V following this decomposition. Let $d' > 0, d' < d$ be such that the map $(X, Y) \rightsquigarrow \exp_G(X) \cdot \exp_G(Y)$ is an analytic diffeomorphism denoted Φ of $(\mathfrak{h}^* \cap B_{d'}(0)) \times (V \cap B_{d'}(0))$ onto an open subset W of G. Let $\tilde{W}$ be an open subset of $\mathcal{G}$ such that $F(\tilde{W}) \subset W$ and such that dF maps the tangent space of any point in $\tilde{W}$ onto $\mathfrak{h}^*$. It follows that the tangent map of $\Lambda = \pi \circ \Phi^{-1} \circ F$ is zero at all points of $\tilde{W}$. This means that Λ is locally constant and since $\lambda(1_\mathcal{G}) = 0$, replacing $\tilde{W}$ by a smaller open subset if necessary, we may assume that $\Lambda = 0$ on $\tilde{W}$. We see thus that (again replacing $\tilde{W}$ by a smaller open set if necessary) that F maps $\tilde{W}$ into $\exp_G(\mathfrak{h}^* \cap B_{d'}(0))$ and the differential map of F as a map into the manifold $\exp_G(\mathfrak{h}^* \cap B_{d'}(0))$ is onto. Replacing $\tilde{W}$ once more by a smaller open neighbourhood of 1 in $\mathcal{G}$, we see that $F(\tilde{W}) \subset H_o$ and is an open subset Ω of H_o in its Lie group topology. Let H_1 be the (open) subgroup of H_o generated by Ω.

5.56. Proof of Theorem 5.53 (Continued). For $i = 1, 2$, let Γ_i' be a countable dense subgroup of G_i whose image in G we denote Γ_i''. Let Γ'' the subgroup of G generated by Γ_1'' and Γ_2''. Let $\Gamma_i = \alpha_i^{-1}(\Gamma'')$. As $\Gamma_i \supset \Gamma_i'$, it is dense in G_i. It is also clear that $\alpha_i(\Gamma_i)$ normalizes $\alpha_{i+1}(\Gamma_{i+1})$. Let Γ be the (countable) subgroup generated by $\alpha_1(\Gamma_1)$ and $\alpha_2(\Gamma_2)$. Let $\mathbf{C}_0 = \{1\}$ and $\mathbf{C}_1 = \{[\alpha_i(g_i), \alpha_i(g_{i+1})] | \ g_i \in G_i, i = 1, 2\}$ and for $n \geq 2$, define inductively $\mathbf{C}_n$ to be $\mathbf{C}_1 \cdot \mathbf{C}_{n-1}$. Then evidently $\cup_{n \in \mathbb{N}} \mathbf{C}_n = H^*$. We will show by induction on n that

$$\mathbf{C}_n \subset \Gamma \cdot H_1. \qquad (**)$$

As $H_1 \subset H^*$, this will prove that $H^* = \Gamma \cdot H_1$. Our theorem would then follow from 5.23. It remains for us to prove $(**)$ above.

5.57. Proof of Theorem 5.53 (Continued). From now on, for $g \in G_i$ we denote $\alpha_i(g)$ simply g. Assume that we have shown that $(**)$ holds for all $n \leq q - 1$. Let $g \in \mathbf{C}_q$. Then $g = u \cdot h$ with $h \in \mathbf{C}_{q-1}$ and $u \in \mathbf{C}_1$. By the induction hypothesis $h = \gamma \cdot c$ with $c \in H_1$. Let $u = [a, b]$ with $a \in G_i$ and

$b \in G_{i+1}$. Then $g = u \cdot h = [a, b] \cdot \gamma \cdot c$. Set $a' = \gamma^{-1} \cdot a \cdot \gamma$ and $b' = \gamma^{-1} \cdot b \cdot \gamma$. We need then to prove that $[a', b'] \in \Gamma \cdot H_1$. Now since Γ_i is dense in G_i, given any open set U_i of G_i, $i = 1, 2$, we can find $\alpha' \in \Gamma_i$ (resp. $\beta' \in \Gamma_{i+1}$) such that $a' = x \cdot \alpha$ (resp. $b' = \beta \cdot y$) with $x \in U_i$ (resp. $y \in U_{i+1}$). We then have

$$[a', b'] = [x \cdot \alpha, \beta \cdot y] = [x \cdot \alpha, \beta] \cdot {}^{\beta}[x \cdot \alpha, y].$$

If the U_i is chosen sufficiently small, ${}^{b}[x \cdot \alpha, y] \in \Omega$, hence in H_1. Thus we need only prove that $[x \cdot \alpha, \beta] \in \Gamma \cdot H_1$. We have

$$[x \cdot \alpha, \beta] = {}^{x}[\alpha, \beta] \cdot [x, \beta] = [\alpha, \beta] \cdot [[a, b]^{-1}, x] \cdot [x, \beta].$$

As $[\alpha, \beta] \in \Gamma$, our contention is proved if $[[a, b]^{-1}, x] \in \Omega$ and $[x, \beta] \in \Omega$ and this is indeed the case, if the U_i are chosen to be sufficiently close to 1. That completes the proof of Theorem 5.53.

From a careful perusal of the proof given above one sees that it yields the following two theorems.

5.58. Theorem. *Let G be an open subgroup in a group of exponential type. Then $H = [G, G]$ has a natural structure of a Lie group such that $(H, i : H \to G)$, i being the inclusion of H in G is a Lie subgroup. Moreover H is an open subgroup in a Lie group H^* where $(H^*, i^* : H^* \to G)$ is a Lie subgroup, H^* is a Lie group of exponential type with $\mathfrak{h} = [\mathfrak{g}, \mathfrak{g}]$ as its Lie algebra and $i = i^*|_H$.*

5.59. Theorem. *Let $(G, i : G \to GL(n, k))$ be a Lie subgroup of $GL(n, k)$ and $\mathfrak{g}$ the corresponding Lie subalgebra of $\mathfrak{gl}(n, k)$. Let $G, G_q, i_q : G_q \to G, q = 1, 2$, be normal Lie subgroups of G with $\mathfrak{g}_q, q = 1, 2$ as the corresponding Lie ideals of $\mathfrak{g}$. Let $\Omega \subset \mathfrak{gl}(n, k)$ be the open subset as in 5.39 and $\Omega_{\mathfrak{g}_q} = \Omega \cap \mathfrak{g}_q$. Assume that for $q = 1, 2$, G_q is generated by $\exp_{\mathfrak{gl}(n,k)}(\Omega_{\mathfrak{g}_q})$. Then $[G_1, G_2]$ is an open subgroup of the Lie subgroup H generated by $\exp_{\mathfrak{gl}(n,k)}(\Omega \cap \mathfrak{h})$ where $\mathfrak{h} = [\mathfrak{g}_1, \mathfrak{g}_2]$ in the Lie group topology on H.*

5.60. Remarks. (1) H and $[G_1, G_2]$ are normal subgroups of G.

(**2**) For a Lie group G, $[G, G]$ need not be closed. Here is an example to illustrate this. Let $\mathcal{H}$ be the (Heisenberg) group of 3×3 upper triangular unipotent matrices over $\mathbb{R}$ and $\mathcal{G} = \mathcal{H} \times \mathbb{R}$. The centre $\mathcal{C}$ of $\mathcal{H}$ is the group $\exp_{GL(3,\mathbb{R})}(\{t \cdot E_{23} | t \in \mathbb{R}\})$ and that of $\mathcal{G}$ is $\mathcal{C} \times \mathbb{R}$. Let α be an irrational number and let Γ be the subgroup $\{\exp_{GL(3,\mathbb{R})}(m \cdot E_{23}, n \cdot \alpha) | m, n \in \mathbb{Z}\}$. Let $G = \mathcal{G}/\Gamma$ and $\pi : \mathcal{G} \to G$ the natural map. Then $[G, G] = \pi(\mathcal{C} \times \{0\})$ is a dense subgroup of the compact group $\pi(\mathcal{C} \times \mathbb{R})$.

(**3**) As has already been noted, when k is archimedean, a Lie group is of exponential type if and only if it is connected. Thus an open subgroup of a Lie group of exponential type is the group itself.

5.61. Recall that the derived series $\{D^i(G)|\ i \in \mathbb{N}\}$ (resp. descending central series) of a group G is the sequence of normal subgroups

$D^0(G) = G, \{D^i(G) = [D^{i-1}(G), D^{i-1}(G)]|\ i \in \mathbb{N}, i > 0\}$
(resp. $C^0(G) = G, \{C^i(G) = [G, D^{i-1}(G)]|\ i \in \mathbb{N}, i > 0\}$).

Similarly for a Lie algebra $\mathfrak{g}$, the derived series $\{D^i(\mathfrak{g})|\ i \in \mathbb{N}\}$ (resp. descending central series) of $\mathfrak{g}$ is the sequence of Lie ideals

$D^0(\mathfrak{g}) = \mathfrak{g}, \{D^i(\mathfrak{g}) = [D^{i-1}(\mathfrak{g}), D^{i-1}(\mathfrak{g})]|\ i \in \mathbb{N}, i > 0\}$
(resp. $C^0(\mathfrak{g}) = \mathfrak{g}, \{C^i(\mathfrak{g}) = [\mathfrak{g}, C^{i-1}(\mathfrak{g})]|\ i \in \mathbb{N}, i > 0\}$).

A simple induction on dim.G shows that we have, as a consequence of Theorem 5.58.

5.62. Corollary to 5.49. *Let G be an open subgroup of a Lie group G' of exponential type. Then for every $i \in \mathbb{N}$, $D^i(G)$ (resp. $C^i(G)$) is in a natural fashion an open subgroup of a Lie subgroup of exponential type corresponding to the ideal $D^i(\mathfrak{g})$ (resp. $C^i(\mathfrak{g})$) of $\mathfrak{g}$.*

5.63. Corollary. *Let G be an open subgroup of a Lie group of exponential type. Then G is solvable (resp. nilpotent) if and only if $\mathfrak{g}$ is solvable (resp. nilpotent).*

This is immediate from the previous corollary. In particular we have

5.64. Corollary. *If G is a connected Lie group (over $\mathbb{R}$ or $\mathbb{C}$), it is solvable (resp. nilpotent) iff its Lie algebra $\mathfrak{g}$ is solvable (resp. nilpotent).*

5.65. We now prove some results on compact Lie groups. Compact Lie groups over archimedean fields differ a great deal from those over non-archimedean fields.

5.66. Theorem. *A compact unipotent Lie group G $(\subset GL_k(V))$ over an archimedean k is trivial.*

5.67. Proof. Let $u \in G$. If $u \neq 1$, $X = ln(u) = \sum_{n \in N}(-1)^n \cdot (u-1)^{n+1}/(n+1)$ is nilpotent. Then we have for $r \in \mathbb{N}$,

$$u^r = \sum_{i=0}^{q} r^i \cdot X^i/i!$$

where $X^q \neq 0$ and $X^{q+1} = 0$. Now since k is archimedean, we have

$$||u^r|| \geq r^q \cdot ||X^q||/q! - \sum_{0 \leq i < q} r^i \cdot ||X^i||$$

and the right hand side in the above inequality evidently tends to ∞ as r tends to ∞. Thus G contains the non-compact closed set $\{u^r|\ r \in \mathbb{N}\}$ and is hence non-compact.

By way of contrast we have seen that the following holds.

5.68. Theorem. *If k is non-archimedean, any Lie group G over k – in particular a unipotent Lie group – has a fundamental system of neighbourhoods which are compact open subgroups of G.*

The group of upper triangular unipotents in $GL(n, k)$ for any $n \geq 2$ is a non-compact unipotent group.

5.69. Theorem. *Assume k to be a p-adic field for some prime $p \in \mathbb{Z}$. Then any Lie group G of exponential type over k is a pro-p group.*

5.70. Proof. By definition of exponential type $G = \exp_G(B_d(0))$ and for all $0 \leq d' \leq d$, $\exp_G(B_{d'}(0))$ is a normal open subgroup of G. In particular, for every $r \in N$, $B_{d/p^r}(0)$ is an open normal subgroup of G. This implies that for every $g \in B_o(d)$, $G^{p^r} \in \exp_G(B_{d/p^r}(0))$. It follows that $G/(\exp_G(B_{d/p^r}(0)))$ is a finite p-group for every $r > 0$. The theorem follows since (as is easy to see) G is the projective limit of the $\{G/\exp_G(B_{d/p^r}(0))| \ r \in \mathbb{N}\}$.

5.71. Theorem. *Assume that k is non-archimedean. Let G be a compact subgroup of $GL_k(V)$, V a finite dimensional vector space of dimension m. Then there is a basis $B = \{e_i| \ 1 \leq i \leq m$ such that $g(\sum_{i=0}^{m} \mathfrak{o}_k \cdot e_i) = \sum_{i=0}^{m} \mathfrak{o}_k \cdot e_i$ for all $g \in G$.*

5.72. Proof. Let B' be a basis of V over k and $L' = \sum_{b \in B} \mathfrak{o}_k \cdot b$. Then as L' is open in V, $G' = \{g \in G| \ g(L') = L'\}$ is an open subgroup of G. Since G is compact, there is a finite subset $\Xi \subset G$ such that $G = \Xi \cdot G'$. Now let $L = \sum_{\xi \in \Xi} \xi(L')$. Then for $g \in G$, and $\xi \in \Xi$, $g \cdot \xi = \xi' \cdot g'$ for some $\xi' \in G'$ and $\xi' \in \Xi$. It follows that $g(\xi(L')) = \xi'(L')$ and hence $g(L) = L$. Now any finitely generated $\mathfrak{o}_k$-submodule of V containing a basis of V over k is free of rank $\dim_k V$. Hence the theorem.

We end this chapter with a result on continuous homomorphisms between Lie groups over non-isomorphic local fields.

5.73. Theorem. *Let k, k' be two local fields and $\hat{\mathbb{Q}}$ (resp. $\hat{\mathbb{Q}}'$) be the closure of $\mathbb{Q}$ in k (resp. k'). Assume that $\hat{\mathbb{Q}}$ is not isomorphic to $\hat{\mathbb{Q}}'$. Let G (resp. G') be a Lie group over k (resp. k') and f a continuous group homomorphism of G in G'. Then the kernel of f is an open subgroup of G and the image is a countable subgroup of G'.*

5.74. Proof. Consider first the case k is archimedean. Then k' is not archimedean. Let G^o be the connected component of 1 in G. Then G^o is open in G. Now G' is totally disconnected and $f(G^o)$ is connected so that $f(G^o) = \{1\}$. Thus kernel G contains G^o and is hence open in G, proving the theorem when k is archimedean.

Suppose next that k and k' are non-archimedean with residue field characteristic p and q respectively. Then G has an open compact subgroup U which is a pro-p group. It follows that the image $f(U) = U'$ is a (compact) pro-p group. By Cartan's theorem U' is a Lie group over $\hat{\mathbb{Q}}'$. Now $q(\neq p)$ is the residue field characteristic of k', U' admits an open subgroup U'' which is a pro-q group; and this can happen only if U'' is trivial. Thus $f^{-1}(U'') \subset \mathrm{kernel}(f)$. It follows that $\mathrm{kernel}(f)$ is open.

Finally consider the case when k is non-archimedean and k' is $\mathbb{R}$ or $\mathbb{C}$. Let p be the residue field characteristic of k and U a compact open pro-p subgroup of G and $U' = f(U)$. Then U' is a compact real Lie group (Cartan's Theorem). Let U'' be the connected component of 1 in U'. Then U'' is an open and closed (compact) subgroup of U'. Let $V = U \cap f^{-1}(U'')$ and V_o the kernel of $f|_V$. Then f induces a bijection of the compact Lie group V/V_o over k on U'', a connected compact Lie group. This bijection is a homeomorphism as V/V_o is compact. The inverse is a homeomorphism of the connected group U'' on V/V_o. It follows that U'' is trivial, proving the theorem in this case as well.

6. Lie Algebras: Theorems of Engel, Lie and Cartan

As we saw in Chapter 5, properties of Lie groups are controlled to a considerable extent by their Lie algebras. So a study of Lie Groups is inevitably tied up with the study of Lie algebras. In this and the next chapter we study the structure of Lie algebras. This is done over an arbitrary field k of characteristic zero – the theorems we prove do not require the assumption that the ground field is a local field. The theorems have implications for Lie groups over k when k is a local field. We draw attention to some of these implications, but this is not done exhaustively. We prove theorems due to Engel and Lie which deal with nilpotent and solvable Lie algebras and a theorem of Cartan's giving a criterion for the solvability of a Lie subalgebra of $\mathfrak{gl}(V)$.

6.1. Notation. In this chapter unless otherwise specified, k will denote an arbitrary field of characteristic 0 (*not necessarily a local field*) and $\bar{k}$ an algebraic closure of k. We have defined Lie algebras and given other basic definitions related to them in Chapter 1 (1.36–1.45). In this chapter, unless otherwise specified, 'algebra' would mean Lie algebra, $\mathfrak{g}$ will denote a Lie algebra and $d_{\mathfrak{g}}$ its dimension. Some of the initial material in this chapter may already be found in Chapter 1.

6.2. Examples. We give below a number of examples of Lie algebras. (**1**) Let V be a vector space over k. Define $[\cdot,\cdot] : V \times V$ by setting $\forall\ X, Y \in V, [X, Y] = 0$. These Lie algebras are *abelian* (see 1.37).

(**2**) Let A be an associative algebra over k. Define for $X, Y \in A$, $[X, Y] = X \cdot Y - Y \cdot X$. Then $(A, [\cdot,\cdot])$ is a Lie algebra (exercise).

(**3**) For a vector space V over k, $A = \text{End}_k(V)$ (denoted also $\mathfrak{gl}_k(V)$ or $\mathfrak{gl}(V)$, if k is known from the context), is a special case of (2). When $\dim_k V$ is finite an ordered basis of V over k gives an isomorphism $\text{End}_k(V)$ on $M(n, k)$.

(**4**) Let $\mathfrak{sl}(V) = \{X \in \mathfrak{gl}(V) \mid \text{trace}(X) = 0\}$. Then $\mathfrak{sl}(V)$ is an ideal in $\mathfrak{gl}(V)$ (exercise).

(**5**) Let A be a non-degenerate alternating bilinear form on a vector space V of finite (even) dimension. Then

$$\mathfrak{sp}(A) = \{T \in \mathrm{End}_k(V)|\ A(T(X), Y) + A(X, T(Y)) = 0\}$$

is a subalgebra (see 1.39) of $\mathfrak{gl}(V)$ (exercise). As any two non-degenerate alternating forms on V are equivalent, the isomorphism class of $\mathfrak{sp}(A)$ is independent of A. $\mathfrak{sp}(A)$ is the *Symplectic* Lie algebra (of A).

(**6**) Let B be a non-degenerate symmetric bilinear form on a finite dimensional vector space V. The *orthogonal* Lie algebra $\mathfrak{o}(B)$ of B is

$$\{T \in \mathrm{End}_k(V)|\ B(T(X), Y) + B(X, T(Y)) = 0\}$$

of $\mathfrak{gl}(V)$ (Exercise: show that it is a subalgebra). In contrast to alternating forms, over a general field, there can be many mutually inequivalent non-degenerate symmetric bilinear forms. For two non-degenerate symmetric forms B, B', the corresponding orthogonal Lie algebras $\mathfrak{o}(B)$ and $\mathfrak{o}(B')$ are isomorphic, if and only if B' is equivalent to $\lambda \cdot B$ for some $\lambda \in k^\times$.

6.3. Representations. We defined the notion of a representation of a group or a Lie algebra and proved some results about them in Chapter 1, V. In 4.46 we defined representations of Lie group over k (when k is a local field) and pointed out that the results in Chapter 1, V, carry over *verbatim* to representations of Lie groups. In the present chapter, we will make free use of the results of Chapter 1, V.

6.4. Theorem. *If a Lie algebra $\mathfrak{g}$ over a local field k admits a faithful representation ρ, then there is a Lie group G over k whose Lie algebra is isomorphic to $\mathfrak{g}$.*

This is an immediate consequence of 5.26.

6.5. Remark. It is known that any Lie algebra over a field of characteristic 0 admits a faithful representation. This result is known as Ado's Theorem. It will be proved in the next chapter.

The proof of the next lemma is left as an exercise to the reader.

6.6. Lemma. *Let ρ, ρ' be F of $\mathfrak{g}$. Let $\{V_i|\ 0 \leq i \leq q\}$ be $\mathfrak{g}$-submodules of V_ρ such that $V_0 = \{0\}, V_q = V_\rho$ and $V_i \subset V_{(i+1)}, V_i \neq V_{(i+1)}$ for $0 \leq i < q$. For $1 \leq i \leq q$, let ρ_i be the representation of $\mathfrak{g}$ on $V_i/V_{(i-1)}$. Then for any $\{X_j|\ 1 \leq j \leq r\} \subset \mathfrak{g}$,*

$$\mathrm{trace}(\rho(X_1) \cdot \rho(X_2) \cdots \rho(X_r)) = \sum_{q=1}^{r} \mathrm{trace}(\rho_i(X_1) \cdot \rho_i(X_2) \cdots \rho_i(X_r))$$

and for $X \in \mathfrak{g}$,

$$\text{trace}(\rho \otimes \rho')(X) = d_{\rho'} \cdot \text{trace}(\rho(X)) + d_\rho \cdot \text{trace}(\rho'(X)).$$

6.7. The Adjoint Representation. Of particular interest is the *adjoint* representation denoted $ad_\mathfrak{g} : \mathfrak{g} \to \mathfrak{gl}(\mathfrak{g})$ (or simply ad if the Lie algebra is known in the context without ambiguity) which is defined as follows: for $X, Y \in \mathfrak{g}$, $ad(X)(Y) = [X, Y]$. That ad is a representation follows from the Jacobi identity. The kernel of the adjoint representation is evidently $\mathfrak{c}(\mathfrak{g})$, the centre of $\mathfrak{g}$. For a representation $\rho : \mathfrak{g} \to \mathfrak{gl}(V)$ of $\mathfrak{g}$, we define a bilinear form B_ρ on $\mathfrak{g}$ by setting for $X, Y \in \mathfrak{g}$, $B_\rho(X, Y) = \text{trace}(\rho(X) \cdot \rho(Y))$. B_ρ is a (symmetric) bilinear form on $\mathfrak{g}$ which is invariant under the adjoint representation, i.e.,

$$B_\rho([X, Y], Z) + B_\rho(Y, [X, Z]) = 0, \ \forall \ X, Y, Z \in \mathfrak{g}.$$

B_{ad}, the *Killing form of* $\mathfrak{g}$ is also denoted $K_\mathfrak{g}$ or $B_\mathfrak{g}$. If $\mathfrak{a}$ is an ideal in $\mathfrak{g}$, it is easily seen that $B_\mathfrak{g}|_\mathfrak{a} = B_\mathfrak{a}$. If B is an $ad_\mathfrak{g}$-invariant bilinear form on $\mathfrak{g}$ and $\mathfrak{a}$ is an ideal in $\mathfrak{g}$, so is $\mathfrak{a}^\perp \stackrel{\text{def}}{=} \{X \in \mathfrak{g}| \ B(X, Y) = 0, \ \forall \ Y \in \mathfrak{a}\}$.

The following result is a restatement of Proposition 1.55.

6.8. Proposition. *Let ρ be a representation of the Lie algebra $\mathfrak{g}$ over k. The following conditions on ρ are equivalent.*
(i) ρ is a direct sum of irreducible representations.
(ii) V_ρ admits a collection of $\{\rho_i | i \in I\}$ of irreducible sub-representations such that $V_\rho = \sum_{i \in I} V_{\rho_i}$.
(iii) If σ is a sub-representation of ρ, there is a sub-representation τ of ρ such that $V_\rho = V_\sigma \oplus V_\tau$.

A representation ρ of $\mathfrak{g}$ is *completely reducible* if it satisfies one of the three equivalent conditions (i)–(iii) in the statement of the proposition.

The next result is again a restatement of Proposition 1.58.

6.9. Proposition. *Let ρ be a representation of $\mathfrak{g}$ and $\bar{\rho}$ its natural extension to $\bar{\mathfrak{g}} = \mathfrak{g} \otimes_k \bar{k}$ on $\bar{V}_\rho = V_r \otimes_k \bar{k}$. Then ρ is completely reducible if and only if $\bar{\rho}$ is ($\bar{k}$ is an algebraic closure of k).*

6.10. Remark. In Chapter 5 we defined the adjoint representation Ad_G of a Lie group on its Lie algebra $\mathfrak{g}$ and showed that $\dot{A}d_G = ad_\mathfrak{g}$.

6.11. Derivations. Recall that a *derivation* $D : \mathfrak{g} \to \mathfrak{g}$ of a Lie algebra $\mathfrak{g}$ is a k-linear map such that for $X, Y \in \mathfrak{g}$, $D([X, Y]) = [D(X), Y] + [X, D(Y)]$ (1.44). The Jacobi identity implies that for $X \in \mathfrak{g}$, $ad(X)$ is a derivation: it is an *inner* derivation. As was already shown, if $D_i, i = 1, 2$ are derivations of $\mathfrak{g}$, so is $D_1 \cdot D_2 - D_2 \cdot D_1$ and the set $\mathfrak{a}(\mathfrak{g})$ of all derivations of $\mathfrak{g}$ is a Lie subalgebra of $\mathfrak{gl}(\mathfrak{g})$. One checks easily that $ad(\mathfrak{g})$ is an ideal in $\mathfrak{a}(\mathfrak{g})$. An ideal $\mathfrak{h}$ in $\mathfrak{g}$ is a *characteristic* ideal, if it is stable under all derivations of $\mathfrak{g}$. It is easily checked that if $\mathfrak{a}, \mathfrak{b}$ are characteristic ideals, so are $\mathfrak{a} \cap \mathfrak{b}$, $\mathfrak{a} + \mathfrak{b}$ and $[\mathfrak{a}, \mathfrak{b}]$.

6.12. Theorem. *Let k be a local field and $\mathfrak{g}$ a Lie algebra over k. Let $A(\mathfrak{g}) \subset GL_k(\mathfrak{g})$ be the group of all (k-linear) Lie algebra automorphisms of $\mathfrak{g}$. Then $A(\mathfrak{g})$ has a natural structure of a Lie group over k and its Lie algebra is isomorphic over k to $\mathfrak{a}(\mathfrak{g})$.*

6.13. This is proved in Chapter 5 (5.38).

6.14. Semi-Direct Products. Suppose now that $\mathfrak{h}$ is a Lie algebra and $\rho : \mathfrak{h} \to \mathfrak{a}(\mathfrak{g})$ is a Lie algebra morphism. Then one defines a natural structure of a Lie algebra on $\mathfrak{h} \oplus \mathfrak{g}$ as follows: on $\mathfrak{g}$ and $\mathfrak{h}$ the bracket operations are those of the respective Lie algebras; for $X \in \mathfrak{h}$ and $Y \in \mathfrak{g}$, $[X, Y] = \rho(X)(Y)$. The fact that $\rho(X)$ is a derivation ensures that the Jacobi identity holds for this bracket operation. It is the *semi-direct product of $\mathfrak{h}$ and $\mathfrak{g}$* and is denoted $\mathfrak{h} \propto_\rho \mathfrak{g}$ or simply $\mathfrak{h} \propto \mathfrak{g}$, if the morphism ρ is known from the context. Evidently, $\mathfrak{h}$ is a subalgebra and $\mathfrak{g}$ is an ideal in $\mathfrak{h} \propto \mathfrak{g}$. Taking $\mathfrak{h}$ to be $\mathfrak{a}(\mathfrak{g})$ and ρ to be the identity $1_{\mathfrak{a}(\mathfrak{g})}$, we have a natural semi-direct product $\mathfrak{a}(\mathfrak{g}) \propto \mathfrak{g}$.

6.15. Theorem. *Let k be a local field and $\mathfrak{h}, \mathfrak{f}$ be Lie algebras over k and $\dot{\rho} : \mathfrak{h} \to \mathfrak{a}(\mathfrak{f})$ a Lie algebra morphism. Suppose that there exist Lie groups H and F with $\mathfrak{h}$ and $\mathfrak{f}$ as their respective Lie algebras, then there is a Lie group G with $\mathfrak{g} = \mathfrak{h} \propto_{\dot{\rho}} \mathfrak{f}$ as its Lie algebra.*

6.16. Proof. Consider first the case when k is archimedean. In this case we may assume that H and F are connected and simply connected. From Theorem 5.34, for each $h \in H$ there is a Lie group automorphism $\tilde{\rho}(h) : F \to F$ which induces $\rho(h)$ on the Lie algebra $\mathfrak{f}$. Moreover that the map $H \times \mathfrak{f} \ni (h, X) \rightsquigarrow \tilde{\rho}(h)(X)$ of $H \times \mathfrak{f}$ in $\mathfrak{f}$ is analytic – one has $\tilde{\rho}(h)(\exp_F(X)) = \exp_F(\rho(h)(X))$. From the analyticity of $\exp_H$ one sees that the map $H \times F \to F$ is analytic. It follows that the semi-direct product of H and F is a Lie group whose Lie algebra is $\mathfrak{h} \propto_\rho \mathfrak{f}$, proving the theorem for archimedean k.

Suppose next that k is non-archimedean. In this case we may assume that H is compact. By 5.34, there is an action of an open compact subgroup H' of H on $\mathfrak{f}$ by Lie algebra automorphisms. As H' is compact, we may assume that it stabilizes the free $\mathfrak{o}_k$-module $L = \sum_{s \in S} \mathfrak{o}_k \cdot s$ for a basis S of $\mathfrak{f}$ (Theorem 5.71). Let

$$\mathfrak{f}_* = \{X \in L | \ [X, A] \in L, \ \forall A \in L\}$$

Then $\mathfrak{f}_*$ is an open set in $\mathfrak{f}$ containing 0. It is clear that if $X, Y \in \mathfrak{f}_*$, $X + Y \in \mathfrak{f}_*$ and $u \cdot X \in \mathfrak{f}_*$ for all $u \in \mathfrak{o}_k$. Further for $X, Y \in \mathfrak{f}_*$ and $B \in L$, $[[X, Y], B] = [X, [Y, B]] + [[X, B], Y] \in L$. Hence $\mathfrak{f}_*$ is a Lie subalgebra over $\mathfrak{o}$ of $\mathfrak{g}$ (contained in the free $\mathfrak{o}$-module L). As L is finitely generated over $\mathfrak{o}_k$, $\mathfrak{f}_*$ is a free module over $\mathfrak{o}_k$ (of rank $\dim._k \mathfrak{f}$) (cf. 1.94). Replacing H' by a compact open subgroup, if need be, we assume that H' stabilizes $\mathfrak{f}_*$. Choosing a basis S of $\mathfrak{f}_*$ over $\mathfrak{o}_k$ and using it (as a basis of $\mathfrak{f}$ over k), we may assume that $\mathfrak{f}_* = B_1(0)$ (refered to the norm given by the basis S). Choose $d > 0, d \leq 1$ such that $\exp_F$ maps $B_d(0)$ diffeomorphically onto a compact open subgroup F_o of F'. Let $H_o = \{h \in H' \mid \dot{\rho}(h)(B_d(0)) = B_d(0)\}$; then H_o is an open subgroup of H. Define for $x = \exp_F(X) \in F_o$ and $h \in H_o$ $\rho(h)(x) = \exp_F(\dot{\rho}(X))$. It is then easy to check that ρ is an anaytic homomorphism of H_o into $\mathrm{Aut}(F_o)$. Let G be the semi-direct product $G = H_o \ltimes_\rho F_o$. Then G has $\mathfrak{h} \ltimes_\rho \mathfrak{f}$ as its Lie algebra.

6.17. Lemma. *Assume that $k = \bar{k}$. Let $D \in \mathrm{End}_k(\mathfrak{g})$ be a derivation. For $\lambda \in k$, let $\mathfrak{g}(\lambda) = \{v \in \mathfrak{g} \mid (D - \lambda)^{d(\mathfrak{g})}(v) = 0\}$. Then for $\lambda, \mu \in k$, $X \in \mathfrak{g}(\lambda)$, $Y \in \mathfrak{g}(\mu)$, $[X, Y] \in \mathfrak{g}(\lambda + \mu)$. ($d(\mathfrak{g}) = $ is $\dim_k \mathfrak{g}$).*

6.18. Proof. The map $t \to t \cdot D$, $t \in k$ can be viewed as a representation δ of the 1-dimensional Lie algebra k on $\mathfrak{g}$. That D is a derivation means that the map $\mathfrak{g} \otimes \mathfrak{g} \to \mathfrak{g}$ given by the bilinear map $(X, Y) \rightsquigarrow [X, Y]$ of $\mathfrak{g} \times \mathfrak{g}$ in $\mathfrak{g}$ is a morphism of $\delta \otimes \delta$ in δ. The eigen-values of $(\delta \otimes \delta)(1)$ $(= D \otimes 1 + 1 \otimes D)$ in $\mathfrak{g}(\lambda) \otimes \mathfrak{g}(\mu)$ are all equal to $\lambda + \mu$ and hence all the eigen-values of D in the image of $\mathfrak{g}(\lambda) \otimes \mathfrak{g}(\mu)$ are $\lambda + \mu$. The lemma follows from this.

6.19. Proposition. *Let $D \in \mathrm{End}_k(\mathfrak{g})$ be a derivation of $\mathfrak{g}$. Then the semisimple part D_s and the nilpotent part D_n of D are derivations.*

6.20. Proof. It suffices to prove this when $k = \bar{k}$. Suppose that $X \in \mathfrak{g}(\lambda)$ and $Y \in \mathfrak{g}(\mu)$ (see Lemma 6.17). $[X, Y] \in \mathfrak{g}(\lambda + \mu)$. Now for $\nu \in k$ and $X \in \mathfrak{g}(\nu)$, one has $D_s(X) = \nu \cdot X$. That D_s is a derivation is immediate from this. Hence so is $D_n = D - D_s$.

6.21. Remarks. (**1**) The Killing form is invariant under all derivations, i.e., $B_\mathfrak{g}(D(X), Y) + B_\mathfrak{g}(X, D(Y)) = 0$, $\forall X, Y \in \mathfrak{g}$ *and* $D \in \mathfrak{a}(\mathfrak{g})$. This is seen as follows: let $\tilde{\mathfrak{g}}$ be the semi-direct product of $\mathfrak{a}(\mathfrak{g})$ and $\mathfrak{g}$. Then since $\mathfrak{g}$ is an ideal in $\tilde{\mathfrak{g}}$, $B_{\tilde{\mathfrak{g}}}|_{\mathfrak{g} \times \mathfrak{g}} = B_\mathfrak{g}$; and $B_{\tilde{\mathfrak{g}}}$ is invariant under $\tilde{\mathfrak{g}}$; also, for $D \in \mathfrak{a}_\mathfrak{g}$ and $X \in \mathfrak{g}$ considered as elements of $\tilde{\mathfrak{g}}$, $[D, X] = D(X)$.

(**2**) When k is a local field the group $A(\mathfrak{g})$ of Lie algebra automorphisms of $\mathfrak{g}$ is a Lie group over k. Its Lie algebra is $\mathfrak{a}(\mathfrak{g})$ (cf. 5.10).

6.22. Derived Series, Solvable Lie Algebras. There are some characteristic ideals (cf. 6.11) in a Lie algebra that arise in a natural fashion. We recall the definition of some of them here. $D^1(\mathfrak{g}) = [\mathfrak{g}, \mathfrak{g}]$ is the *commutator* ideal of $\mathfrak{g}$; evidently $\mathfrak{g}/D^1(\mathfrak{g})$ is abelian and any ideal $\mathfrak{h}$ such that $\mathfrak{g}/\mathfrak{h}$ is abelian contains $D^1(\mathfrak{g})$. The *derived series* of $\mathfrak{g}$ is the decreasing sequence $\{D^i(\mathfrak{g})\}_{i \in N}$ of ideals defined inductively as follows: $D^0(\mathfrak{g}) = \mathfrak{g}$; $D^1(\mathfrak{g})$ has been defined above; and for $i > 1$, $D^i(\mathfrak{g}) = [D^{(i-1)}(\mathfrak{g}), D^{(i-1)}(\mathfrak{g})]$. Evidently, for every integer $i \geq 0$, $D^i(\mathfrak{g})$ is a characteristic ideal in $\mathfrak{g}$ (6.11) and $D^i(\mathfrak{g})/D^{(i+1)}(\mathfrak{g})$ is abelian. Note that $D^i(D^j \mathfrak{g}) = D^{(i+j)}(\mathfrak{g})$. As $\mathfrak{g}$ is finite dimensional, there is a minimal integer $ld(\mathfrak{g}) \geq 0$ such that $D^q(\mathfrak{g}) = D^{ld(\mathfrak{g})}$ for all $q \geq ld(\mathfrak{g})$. $ld(\mathfrak{g})$ is the *length of the derived series of* $\mathfrak{g}$. A Lie algebra $\mathfrak{g}$ is *solvable* if $D^{ld(\mathfrak{g})}\mathfrak{g} = 0$ (equivalently $D^q(\mathfrak{g})$ is 0 for some $q \geq 0$).

We record for future use the following simple fact as a

6.23. Lemma. *Any linear subspace* $E \supset [\mathfrak{g}, \mathfrak{g}]$ *of* $\mathfrak{g}$ *is an ideal. If* $[\mathfrak{g}, \mathfrak{g}] \neq \mathfrak{g}$ *(e.g.,* $\mathfrak{g}$ *is solvable),* $\mathfrak{g}$ *admits an ideal of co-dimension 1.*

6.24. Corollary. *If* k *is a local field, then for any solvable Lie algebra* $\mathfrak{g}$ *over* k*, there is a Lie group* G *with* $\mathfrak{g}$ *as its Lie algebra.*

6.25. Proof. We argue by induction on $d_\mathfrak{g} = \dim_k \mathfrak{g}$. When $d_\mathfrak{g} = 1$, G can be taken as k. Assume theorem proved for $d_\mathfrak{g} < n$. Assume that $d_\mathfrak{g} = n$. By the lemma above, $\mathfrak{g}$ has an ideal $\mathfrak{b}$ of co-dimension 1. If X is any element of $\mathfrak{g} - \mathfrak{b}$, $\mathfrak{g}$ is a semi-direct product of the Lie subalgebra $k \cdot X$ and the ideal $\mathfrak{b}$. The corollary now follows from 6.15.

6.26. Subalgebras and Quotient Algebras. It is clear that if $\mathfrak{h}$ is a subalgebra of $\mathfrak{g}$, then $D^q(\mathfrak{h}) \subset D^q(\mathfrak{g})$. Thus if $\mathfrak{g}$ is solvable, every subalgebra is solvable. Next suppose that $\mathfrak{g}$ is solvable and $\mathfrak{s}$ is an ideal in $\mathfrak{g}$. Let $\mathfrak{h} = \mathfrak{g}/\mathfrak{s}$ and $\pi : \mathfrak{g} \to \mathfrak{h}$ the natural map. Then $\pi(D^q(\mathfrak{g})) = D^q(\mathfrak{h})$; it follows that if $\mathfrak{g}$ is solvable, so is $\mathfrak{h} = \mathfrak{g}/\mathfrak{s}$. Conversely if the ideal $\mathfrak{s}$ in $\mathfrak{g}$ is solvable and the quotient $\mathfrak{h} = \mathfrak{g}/\mathfrak{s}$ is solvable, $\mathfrak{g}$ is solvable. This is seen as follows: if p is such that $D^p(\mathfrak{h}) = \{0\}$, $D^p(\mathfrak{g}) \subset \pi^{-1}(D^p(\mathfrak{h})) = \mathfrak{s}$; let q be such that $D^q(\mathfrak{s}) = \{0\}$; then $D^{(p+q)}(\mathfrak{g}) = D^q(D^p(\mathfrak{g})) \subset D^q(\mathfrak{s}) = \{0\}$.

6.27. Proposition. *The Lie subalgebra* $\mathfrak{s}(n) = \{A \in M(n,k)|\ A_{i,j} = 0 \text{ for } i < j\}$ *of* $M(n,k)$ *is solvable.*

The proof is left as an exercise to the reader.

6.28. Proposition. *The following conditions on* $\mathfrak{g}$ *are equivalent:*
(i) $\mathfrak{g}$ *is solvable.*
(ii) $\mathfrak{g}$ *admits a decreasing sequence of ideals* $\mathfrak{a}_i$, $0 \leq i \leq r$ *such that* $\mathfrak{a}_i/\mathfrak{a}_{(i+1)}$ *is abelian,* $\mathfrak{a}_r = \{0\}$ *and* $\mathfrak{a}_0 = \mathfrak{g}$.
(iii) $\mathfrak{g}$ *admits a decreasing sequence of subalgebras* $\mathfrak{a}_i$, $0 \leq i \leq r$ *with* $\mathfrak{a}_0 = \mathfrak{g}$, $\mathfrak{a}_r = 0$ *such that* $\mathfrak{a}_{(i+1)}$ *is an ideal in* $\mathfrak{a}_i$ *and* $\mathfrak{a}_i/\mathfrak{a}_{(i+1)}$ *is abelian.*

6.29. Proof. (i) $\Rightarrow$ (ii): $\{D^i(\mathfrak{g})|\ 1 \leq i \leq d_\mathfrak{g}\}$ is a sequence of ideals satisfying the requirement in (ii) with $r = ld_\mathfrak{g}$. (ii) $\Rightarrow$ (iii) is obvious. To prove that (iii) $\Rightarrow$ (i), we argue by induction on r, the number of subalgebras in the sequence. When $r = 1$, $\mathfrak{g}$ is abelian and hence solvable. When $r > 1$, (iii) holds for $\mathfrak{g}' = \mathfrak{a}_1$ in place of $\mathfrak{g}$ and $\{\mathfrak{a}_{(i+1)}|\ 0 \leq i \leq (r-1)\}$ in place of $\{\mathfrak{a}_i|\ 0 \leq i \leq r\}$. By induction hypothesis, $\mathfrak{a}_1$ is solvable. As $\mathfrak{g}/\mathfrak{g}'$ is abelian, $\mathfrak{g}$ is solvable (6.26).

6.30. Proposition. *If $\mathfrak{a}$ and $\mathfrak{b}$ are solvable ideals in $\mathfrak{g}$ then so is $\mathfrak{a} + \mathfrak{b}$. $\mathfrak{g}$ has a unique maximal solvable ideal $\mathfrak{r}(\mathfrak{g})$.*

The maximal solvable ideal $\mathfrak{r}(\mathfrak{g})$ of $\mathfrak{g}$ is the *(solvable) radical of* $\mathfrak{g}$.

6.31. Proof. $(\mathfrak{a}+\mathfrak{b})/\mathfrak{b} \simeq \mathfrak{a}/(\mathfrak{a} \cap \mathfrak{b})$ and is therefore solvable. As $\mathfrak{b}$ is solvable, so is $\mathfrak{a} + \mathfrak{b}$ (6.26). The second assertion follows from the first.

6.32. Descending Central Series, Nilpotent Algebras. The *Descending Central Series* $\{DC^i(\mathfrak{g})\}_{i \in \mathbb{N}}$ of a Lie algebra $\mathfrak{g}$ is the sequence of (characteristic) ideals defined as follows: $DC^0(\mathfrak{g}) = \mathfrak{g}$, $DC^1(\mathfrak{g}) = D^1(\mathfrak{g})$ and for $i > 1$, $DC^i(\mathfrak{g}) = [\mathfrak{g}, DC^{(i-1)}(\mathfrak{g})]$. Evidently, there is a smallest non-negative integer $lc_\mathfrak{g}$ (also denoted $lc(\mathfrak{g})$), the *length of the descending central series*, such that $DC^q(\mathfrak{g}) = DC^{lc_\mathfrak{g}}(\mathfrak{g})$ for all $q \geq lc_\mathfrak{g}$. It is clear that $DC^i(\mathfrak{g})/DC^{(i+1)}(\mathfrak{g})$ is central in $\mathfrak{g}/DC^{(i+1)}(\mathfrak{g})$ for all $i \in \mathbb{N}$. A Lie algebra $\mathfrak{g}$ is *nilpotent*, if and only if $DC^q(\mathfrak{g}) = 0$ for all large q (equivalently, $D^{lc_\mathfrak{g}}(\mathfrak{g}) = 0$). An abelian Lie algebra is evidently nilpotent. The proposition below is immediate from the definitions.

6.33. Proposition. *Suppose that $\mathfrak{g}$ is nilpotent. Then it is solvable. If $\mathfrak{g} \neq \{0\}$, $\mathfrak{c}(\mathfrak{g}) \neq \{0\}$. Any subalgebra or quotient of $\mathfrak{g}$ is nilpotent.*

6.34. Proof. We may assume that $d_\mathfrak{g} \geq 1$. The first assertion holds since $DC^q(\mathfrak{g} \supset D^q(\mathfrak{g}))$. The last non-zero term of the descending central series is a central ideal; hence the second assertion. Let $\mathfrak{h}$ be a subalgebra (resp. an ideal) in $\mathfrak{g}$. Then $DC^q(\mathfrak{h}) \subset DC^q(\mathfrak{g})$ (resp. $DC^q(\mathfrak{g}/\mathfrak{h})$ is the image of $DC^q(\mathfrak{g})$ in $\mathfrak{g}/\mathfrak{h}$). The third assertion is immediate from this.

6.35. Proposition. *Set $\mathfrak{s}(n) = \{\{a_{i,j}\}_{1 \leq i,j \leq n} \in M(n,k)|\, a_{i,j} = 0 \text{ for } i > j\}$ and $\mathfrak{n}(n) = \{\{a_{i,j}\}_{1 \leq i,j \leq n} \in \mathfrak{s}(n)|\, a_{i,i} = 0, \forall i\}$. Then $\mathfrak{n}(n)$ is an ideal in $\mathfrak{s}(n)$ and is a nilpotent algebra. $\mathfrak{s}(n)$ is not nilpotent for $n > 1$.*

One need only prove that $\mathfrak{n}(q) = \{\{a_{i,j}\}_{1 \leq i,j \leq n} \in \mathfrak{n}(n)|\, a_{i,j} = 0 \text{ for } i-q > j\}$ is the descending central series of $\mathfrak{n}(n)$; this is left as an exercise.

6.36. Proposition. *Let $\mathfrak{c}$ be a central ideal in a Lie algebra $\mathfrak{g}$. Then $\mathfrak{g}$ is nilpotent if and only if $\mathfrak{h} = \mathfrak{g}/\mathfrak{c}$ is nilpotent.*

Let $\pi : \mathfrak{g} \to \mathfrak{h}$ be the natural map. Then $DC^q(\mathfrak{g}) \subset \pi^{-1}(DC^q(\mathfrak{h}))$. The proposition follows.

6.37. Remark. Let $\mathfrak{a}(k)(\simeq k)$ denote the Lie algebra of derivations of the abelian Lie algebra k. Let $\mathfrak{g}$ be the semi-direct product of $\mathfrak{a}(k)$ and k. Then k is an abelian (and hence nilpotent) ideal in $\mathfrak{g}$. The quotient $\mathfrak{g}/k \simeq \mathfrak{a}(k) \simeq k$ is again nilpotent. However $\mathfrak{g}$ is not nilpotent. In fact one has $DC^q(\mathfrak{g}) = k$ for all $q \geq 1$.

6.38. Proposition. *Let $\mathfrak{g}$ be a nilpotent Lie algebra. Then there is an ordered basis B of $\mathfrak{g}$ such that all the endomorphisms in $\mathrm{ad}_{\mathfrak{g}}(\mathfrak{g})$ referred to the basis B are upper triangular nilpotent matrices.*

B need only be chosen such that $B \cap DC^q(\mathfrak{g})$ is a basis of $DC^q(\mathfrak{g})$ for every q and the elements of $B - DC^{(q)}(\mathfrak{g})$ follow those in $B \cap DC^q(G)$ in the order.

6.39. Proposition. *Let $\mathfrak{g}$ be nilpotent and $\mathfrak{h} \neq \mathfrak{g}$ a subalgebra. Then the normalizer $\mathfrak{n}(\mathfrak{h}) = \{X \in \mathfrak{g} | [X, \mathfrak{h}] \subset \mathfrak{h}\}$ of $\mathfrak{h}$ contains $\mathfrak{h}$ strictly.*

6.40. Proof. We argue by induction on $d_{\mathfrak{g}} = \dim_k \mathfrak{g}$. The assertion is trivially true when $d_{\mathfrak{g}} = 1$. Assume the assertion proved for $d_{\mathfrak{g}} < n$ and let $d_{\mathfrak{g}} = n$. As $\mathfrak{g}$ is nilpotent, $\mathfrak{c}(\mathfrak{g}) \neq \{0\}$ (see 6.33). Now $\mathfrak{c}(\mathfrak{g}) \subset \mathfrak{n}[\mathfrak{h}]$; so the proposition holds if $\mathfrak{c}(\mathfrak{g})$ is not contained in $\mathfrak{h}$. If $\mathfrak{h} \supset \mathfrak{c}(\mathfrak{g})$, let $\bar{\mathfrak{g}} = \mathfrak{g}/\mathfrak{c}(\mathfrak{g})$ and $\bar{\mathfrak{h}}$ the image of $\mathfrak{h}$ under the natural map $\pi : \mathfrak{g} \to \bar{\mathfrak{g}}$. Now $\mathfrak{n}[\bar{\mathfrak{h}}]$ contains $\bar{\mathfrak{h}}$ strictly (induction hypothesis) and hence $\mathfrak{n}[\mathfrak{h}] = \pi^{-1}(\mathfrak{n}[\bar{\mathfrak{h}}])$ contains $\mathfrak{h}$ strictly. Hence the proposition.

6.41. The *ascending central series* $\{AC_i(\mathfrak{g})\}_{i \in \mathbb{N}}$ of $\mathfrak{g}$ is the increasing sequence of ideals defined as follows: $AC_0(\mathfrak{g}) = \mathfrak{c}(\mathfrak{g})$, the centre of $\mathfrak{g}$ and inductively for $i > 0$, $AC_i(\mathfrak{g})$ is the inverse image of the centre of $\mathfrak{g}/AC_{(i-1)}(\mathfrak{g})$ in $\mathfrak{g}$ under the natural map $\mathfrak{g} \to \mathfrak{g}/AC_{(i-1)}(\mathfrak{g})$.

We will now prove.

6.42. Proposition. *The following conditions on $\mathfrak{g}$ are equivalent:*
(i) *$\mathfrak{g}$ is nilpotent.*
(ii) *There is a sequence $\{\mathfrak{g}_i | 0 \leq i \leq r\}$ of ideals in $\mathfrak{g}$ such that $\mathfrak{g}_0 = \mathfrak{g}$, $\mathfrak{g}_r = \{0\}$, $\mathfrak{g}_{(i+1)} \subset \mathfrak{g}_i$ and $[\mathfrak{g}, \mathfrak{g}_i] \subset \mathfrak{g}_{(i+1)}$.*
(iii) *There is an increasing sequence $\{\mathfrak{a}_i\}_{0 \leq i \leq r}$ of ideals in $\mathfrak{g}$ with $\mathfrak{a}_0 = \{0\}$, $\mathfrak{a}_r = \mathfrak{g}$, and $\mathfrak{a}_{(i+1)}/\mathfrak{a}_i \subset \mathfrak{c}(\mathfrak{g}/\mathfrak{a}_i)$, the centre of $\mathfrak{g}/\mathfrak{a}_i$.*
(iv) *$AC^i(\mathfrak{g}) = \mathfrak{g}$ for all large i.*

6.43. Proof. (i) $\Rightarrow$ (ii): Let $r = lc_{\mathfrak{g}}$. Then $\mathfrak{g}_i = DC^i(\mathfrak{g})$ satisfy (ii).
(ii) $\Rightarrow$ (iii): Let $\mathfrak{g}_i$ be as in (2). Then $\mathfrak{a}_i = \mathfrak{g}_{(r-i)}$ satisfy (iii).
(iii) $\Rightarrow$ (iv): A simple induction on i shows that $AC_i(\mathfrak{g})$ contains $\mathfrak{a}_i$.
(iv) $\Rightarrow$ (i): We will argue by induction on $d_{\mathfrak{g}}$. When $d_{\mathfrak{g}} \leq 1$, $\mathfrak{g}$ is abelian, hence nilpotent. Assume the implication proved for $d_{\mathfrak{g}} < n$ and assume $d_{\mathfrak{g}} = n > 0$. Condition (iv) implies that $\mathfrak{c}(\mathfrak{g}) = AC_0(\mathfrak{g}) \neq \{0\}$. Set $\mathfrak{g}' = \mathfrak{g}/\mathfrak{c}(\mathfrak{g})$. Then $\mathfrak{g}'$ satisfies (iv). By the induction hypothesis ($d_{\mathfrak{g}'} < n$), $\mathfrak{g}'$ is nilpotent and hence so is $\mathfrak{g}$ (see 6.36).

6.44. Proposition. *If $\mathfrak{a}$ and $\mathfrak{b}$ are nilpotent ideals in $\mathfrak{g}$, so is $\mathfrak{a} + \mathfrak{b}$. $\mathfrak{g}$ admits a unique maximal nilpotent ideal $\mathfrak{n}(\mathfrak{g})$.*

The unique maximal nilpotent ideal $\mathfrak{n}(\mathfrak{g})$ of $\mathfrak{g}$ is the *nil-radical* of $\mathfrak{g}$.

6.45. Proof. We assume, as we may, that $\mathfrak{g} = \mathfrak{a} + \mathfrak{b}$. We argue by induction on $\dim_k \mathfrak{g}$. If $\mathfrak{b}$ or $\mathfrak{a}$ is $\{0\}$ there is nothing to prove, so assume that both $\mathfrak{a}$ and $\mathfrak{b}$ are non-zero. Assume the proposition to be true for $d_\mathfrak{g} < n$ and let $d_\mathfrak{g} = n$. Let $\mathfrak{c}(\mathfrak{b})$ $(\neq \{0\})$ be the centre of $\mathfrak{b}$ (see 6.33). Set $DC^{-1}(\mathfrak{a}) = \mathfrak{g}$ and let q be the largest integer ≥ -1 such that $\mathfrak{u} = DC^q(\mathfrak{a}) \cap \mathfrak{c}(\mathfrak{b}) \neq \{0\}$. Then $[\mathfrak{a}, DC^q(\mathfrak{a}) \cap \mathfrak{c}] \subset DC^{(q+1)}(\mathfrak{a}) \cap \mathfrak{c} = \{0\}$. Thus $\mathfrak{u}$ is a non-zero central ideal in $\mathfrak{g}$. Let $\pi : \mathfrak{g} \to \mathfrak{h} = \mathfrak{g}/\mathfrak{c}$ be the natural map. Since $\pi(\mathfrak{a})$ and $\pi(\mathfrak{b})$ are nilpotent and $d_\mathfrak{h} < n$, by induction hypothesis $\mathfrak{h}$ is nilpotent. It follows from 6.36 that $\mathfrak{g}$ is nilpotent.

6.46. Proposition. *Let $\mathfrak{n}(\mathfrak{g})$ be the nil-radical* (see 6.44) *of the Lie algebra $\mathfrak{g}$. Then $ad_\mathfrak{g}(X)$ is nilpotent for every $X \in \mathfrak{n}(\mathfrak{g})$.*

6.47. Proof. This follows from the fact that for $X \in \mathfrak{n}(\mathfrak{g})$, $ad_\mathfrak{g}(X)(\mathfrak{g}) \subset \mathfrak{n}(\mathfrak{g})$ and $ad(X)(DC^q(\mathfrak{n}(\mathfrak{g}))) \subset DC^{(q+1)}(\mathfrak{n}(\mathfrak{g}))$ for all $q \in \mathbb{N}$.

We now prove a central result about Lie algebras which is a kind of converse to 6.46.

6.48. Engel's Theorem. *Let V be a finite dimensional vector space. Let $\mathfrak{n}$ be a Lie subalgebra of $\mathrm{End}_k(V)$ such that every $X \in \mathfrak{n}$ is nilpotent. Then there is a $v \neq 0$ in V such that $X(v_o) = 0$, $\forall X \in \mathfrak{n}$.*

6.49. Proof. We will argue by induction on $\dim_k \mathfrak{n}$. Suppose that the theorem is true for all Lie algebras of nilpotent endomorphisms of dimension less than q. Let $\dim_k \mathfrak{n} = q$. We observe first that if X in $\mathrm{End}_k(V)$ is nilpotent, then $ad_{\mathrm{End}_k(V)}(X)$ (as an element of $\mathrm{End}_k(\mathrm{End}_k(V))$) is also nilpotent. It follows that for all $X \in \mathfrak{n}$, $ad_\mathfrak{n}(X) : \mathfrak{n} \to \mathfrak{n}$ is nilpotent. Let $\mathcal{N}$ be the collection of all proper subalgebras of $\mathfrak{n}$. If $\mathcal{N} = \phi$, $\dim_k \mathfrak{n} = 0$ or 1 and the theorem obviously holds. Assume then that $\mathcal{N}$ is non-empty. If $\mathfrak{h} \in \mathcal{N}$, $ad_\mathfrak{h}$ is a sub-representation of $ad_\mathfrak{n}|_\mathfrak{h}$ and thus we have the quotient representation $\tau = ad_\mathfrak{n}|_\mathfrak{h}/ad_\mathfrak{h}$ of $\mathfrak{h}$ on $W = \mathfrak{n}/\mathfrak{h}$. Now $\tau(X)$ is nilpotent for every $X \in \mathfrak{h}$. Since $\dim_k \mathfrak{h} < q$, by the induction hypothesis, there is a vector $0 \neq z \in W$ such that $\tau(X)(z) = 0$. Let $Z \in \mathfrak{n}$ map to z under the natural map $\mathfrak{n} \to W$; then $[Z, \mathfrak{h}] \subset \mathfrak{h}$ so that $\mathfrak{h}' = (k \cdot Z) + \mathfrak{h}$ is a Lie subalgebra of $\mathfrak{n}$ in which $\mathfrak{h}$ is an ideal. Since $z \neq 0$, $\dim_k \mathfrak{h}' = \dim_k \mathfrak{h} + 1$. Suppose now that $\mathfrak{h}$ is a maximal element in $\mathcal{N}$ (for the partial order of inclusion). Then clearly $\mathfrak{h}' = \mathfrak{n}$. Thus we see that $\mathfrak{n}$ admits an ideal $\mathfrak{h}$ of co-dimension 1 so that $\mathfrak{n} = \mathfrak{h} \oplus (k \cdot Z)$ for some non-zero $Z \in \mathfrak{n}$. Let $V' = \{v \in V | X \cdot v = 0, \forall X \in \mathfrak{h}\}$. By the induction hypothesis, $V' \neq \{0\}$. Since $\mathfrak{h}$ is an ideal in $\mathfrak{n}$, V' is stable under all of $\mathfrak{n}$. Now Z being nilpotent, there is a non-zero vector $v_o \in V'$ such that $Z(v_o) = 0$. Evidently, $\mathfrak{n}(v_o) = \{0\}$. Hence the theorem.

The following corollaries are all easily deduced from the theorem.

6.50. Corollary. *Let ρ be a an irreducible representation of $\mathfrak{g}$. If $\rho(X)$ is nilpotent for every $X \in \mathfrak{g}$, then $\rho(\mathfrak{g}) = \{0\}$.*

6.51. Corollary. *Let ρ be a representation of a Lie algebra $\mathfrak{g}$ such that $\rho(\mathfrak{g})$ consists entirely of nilpotent matrices. Then there are vector subspaces $\{V_i | 1 \leq i \leq q\}$ of V with $V_1 = V_\rho$, $V_q = \{0\}$, $V_i \supset V_{(i+1)}, V_i \neq V_{(i+1)}$ for $1 \leq i \leq (q-1)$, such that $\rho(\mathfrak{g})(V_i) \subset V_{(i+1)}$ for $1 \leq i \leq (q-1)$. The V_i can be chosen such that $\dim(V_i/V_{(i+1)}) = 1$.*

6.52. Corollary. *Let ρ be as in 6.51. Then there is an ordered basis $B = \{e_i | 1 \leq i \leq n\}$ of V_ρ such that the matrices of all the endomorphisms in $\rho(\mathfrak{g})$ with respect to the basis B are upper triangular and nilpotent.*

6.53. Corollary. *A Lie algebra $\mathfrak{g}$ is nilpotent if and only if $ad(X)$ is nilpotent for every $X \in \mathfrak{g}$.*

Note that the kernel of $ad_\mathfrak{g}$ is the centre of $\mathfrak{g}$.

6.54. We will now show that Engel's theorem and its corollaries have implications for Lie groups (over local fields). An element $T \in GL_k(V)$, V, a vector space of dimension n, is *unipotent*, if all its eigen-values are 1; equivalently if $T - 1$ is nilpotent. A subgroup G of $GL_k(V)$ is *unipotent*, if every $g \in G$ is unipotent. The group $N = \{g \in GL(n, k) | g_{i,j} = \delta_{i,j} \ for \ n \geq i \geq j \geq 1\}$ is a unipotent group. Then we have the following.

6.55. Theorem. *Let V be a vector space of dimension $n > 0$. Let U be a unipotent subgroup of $GL_k(V)$. Then the subspace $V_1 = \{v \in V | g(v) = v, \ \forall g \in U\}$ is non-zero. V admits vector subspaces $V_i, 0 \leq i \leq n$, $n = \dim_k V$ such that $V_i \subset V_{i+1}$ for $0 \leq i < n$, $V_0 = \{0\}, V_n = V$, $\dim_k.(V_i/V_{i-1}) = 1$, V_i is U-stable and U acts trivially on V_i/V_{i+1}.*

6.56. Proof of 6.55. The second assertion is immediate from the first. So we need only prove the first assertion. We observe that we can assume that U is finitely generated. This is seen as follows: Let $\mathcal{F}$ be the set of all finitely generated subgroups of U. For $F \in \mathcal{F}$, let $n(F)$ be the dimension of $V^F = \{v \in V | \gamma(v) = v, \ \forall g \in F\}$. Let $F_o \in \mathcal{F}$ be such that $n(F_o) \leq n(F)$, $\forall F \in \mathcal{F}$. Since we have assumed that the theorem holds for finitely generated groups, we see that $n(F_o) \geq 1$. We claim that $V^{F_o} = V^U$. If $x \in U$ and F' is the subgroup of U generated by F_o and x, $F' \in \mathcal{F}$. Now $V^{F'} \subset V^{F_o}$; and by the minimality of $n(F_o)$, $V^{F'} = V^{F_o}$. Thus $\{0\} \neq V^U (= V^{F_o})$. From now on, we assume that U is generated by a finite set $S = \{s_i | 1 \leq i \leq q\}$.

6.57. Proof of 6.55 (Continued). Replacing k by the field generated by the matrix entries of the $\{s_i|\ 1 \le i \le q\}$, we may assume that k is finitely generated over $\mathbb{Q}$. It is then a finite extension of a purely transcendental extension of finite transcendence degree over $\mathbb{Q}$. Fix a prime p. As $\mathbb{Q}_p$ has infinite transcendence degree over $\mathbb{Q}$, k imbeds in a p-adic field K. Let $V_K = V \otimes_k K$. We then have a natural extension of the action of U on V to one on V_K and $V_K^U = V^U \otimes_k K$. Thus to prove the theorem, we may assume that k is a p-adic field.

6.58. Proof of 6.55 (Continued). We assume that k is a p-adic field and U is generated by a finite set $S = \{s_i|\ 1 \le i \le q\} \subset GL_k(V)$. Now V may be regarded as a vector space over $\mathbb{Q}_p$ and since $GL_k(V) \subset GL_{\mathbb{Q}_p}(V)$, we may assume that $k = \mathbb{Q}_p$. The closure $\bar{U}$ of U in $GL_k(V)$ is then a (unipotent) Lie group (Cartan's Theorem). Let $\mathfrak{u}$ be its Lie algebra. Then $\mathfrak{u}$ consists entirely of nilpotent matrices: if $X \in \mathfrak{u}$, $\exp_{GL_k(V)}(t \cdot X) \in \bar{U}$, $\forall t \in k$ near $0\}$. By Engel's theorem there is a $v \in V$, $v \ne 0$ such that $\mathfrak{u}(v) = 0$ and hence $\exp_{GL_k(V)}(\mathfrak{u})(v) = v$. It suffices thus to show that every $u \in U$ is contained in $\exp_{GL_k(V)}(\mathfrak{u})$. Suppose that $g \in U$. Set $X = \ln(1+g-1) = \sum_{r \in N}(-1)^r \cdot (g-1)^{2 \cdot r+1}/r$ (note that the summation is finite since $(g-1)^q = 0$ for $q \ge \dim_k V$); then X is nilpotent and for $n \in \mathbb{Z}$, $\exp_{GL_k(V)}(n \cdot X) = g^n \in U$. Now $\mathbb{Z}$ is dense in $\mathbb{Z}_p$, it follows that $\exp_{GL_k(V)}(\mathbb{Z}_p \cdot X) \subset \bar{U}$ and hence $X \in \mathfrak{u}$ and $g = \exp_{GL_k(V)}(X)$ fixes v.

6.59. Corollary. *If ρ is an irreducible representation of a Lie group G such that $\rho(g)$ is unipotent for every $g \in G$, $\rho(G) = \{1\}$.*

6.60. Corollary. *Any unipotent subgroup U of $GL(n,k)$ is nilpotent.*

U is contained in the group N of all upper triangular unipotents (with respect to a suitable basis of k^n) which is a nilpotent group.

6.61. Corollary. *Let ρ be a representation of the Lie group G such that $\rho(G)$ consists entirely of unipotent automorphisms of V_ρ. Then there is a basis B of V_ρ such that, referred to the basis B, $\rho(g)$ is an upper triangular (unipotent) matrix for every $g \in G$.*

The next lemma has a crucial role to play in Lie theory.

6.62. Lemma (E B Dynkin). *Let $\mathfrak{g}$ be a Lie algebra and $\mathfrak{a}$ an ideal in it. Let ρ be a representation of $\mathfrak{g}$ and λ a k-linear map of $\mathfrak{a}$ in k. Then the subspace $W = \{v \in V_\rho|\ \rho(X)(v) = \lambda(X) \cdot v, \ \forall X \in \mathfrak{a}\}$ is stable under $\rho(\mathfrak{g})$.*

6.63. Proof of Lemma. If $W = \{0\}$, there is nothing to prove. Let $w_0 \in W$, $w_0 \ne 0$. Let $0 \ne X \in \mathfrak{g}$ and for $n \in \mathbb{N}$, set $w_n = \rho(X)^n(w_0)$. Then it is easily seen by induction on n (using the fact that $\rho(A) \cdot \rho(X) = \rho(X) \cdot \rho(A) + \rho([A, X])$), that we have for $A \in \mathfrak{a}$,

$$\rho(A)(\rho(X)^n(w_0)) = \sum_{p=0}^{n} (-1)^p \binom{n}{p} \lambda((ad(X))^p(A)) \cdot \rho(X)^{(n-p)}(w_0). \quad (*)$$

Let d be the maximal integer such that $\{w_i | \ 0 \le i \le (d-1)\}$ are linearly independent over k. Let W' be the (d-dimensional) vector subspace spanned by $\{w_i | \ 0 \le i \le (d-1)\}$. Evidently, W' is stable under $\rho(X)$. From $(*)$ above, we see that W' is stable under $\mathfrak{a}$ as well. Let σ denote the representation of the Lie algebra $(k \cdot X) + \mathfrak{a}$ on W'. Then from $(*)$ again we see that $\text{Trace}(\sigma(A)) = d \cdot \lambda(A)$, $\forall A \in \mathfrak{a}$. It follows that $0 = \text{Trace}(\sigma([X, A])) = d \cdot \lambda([X, A])$ and hence $\lambda([X, A]) = 0$ for all $A \in \mathfrak{a}$. Appealing again to $(*)$, we conclude that $\rho(A)(\rho(X)(w)) = \lambda(A) \cdot \rho(X)(w)$ for all $A \in \mathfrak{a}$ and $w \in W$ proving the lemma.

6.64. Theorem (Sophus Lie). *Assume that k is algebraically closed (i.e., $k = \bar{k}$) and $\mathfrak{g}$ is solvable. Then for any representation ρ of $\mathfrak{g}$, there is a vector $0 \ne v \in V_\rho$ and a linear form $\lambda : \mathfrak{g} \to k$ such that $\rho(X)(v) = \lambda(X) \cdot v$ for all $X \in \mathfrak{g}$.*

6.65. Proof. We argue by induction on $\dim_k \mathfrak{g}$. When $\dim_k \mathfrak{g} = 1$, it is just the statement that over an algebraically closed field every endomorphism of a vector space has a non-zero eigen-vector. Assume the theorem proved for all Lie algebras of dimension less than n ($n \ge 2$). Let $\mathfrak{g}$ be of dimension n. Since $\mathfrak{g}$ is solvable it admits an ideal $\mathfrak{a}$ of co-dimension 1 (see 6.23). By the induction hypothesis, there is a linear form λ on $\mathfrak{a}$ such that $W = \{V \in V_\rho | \ \rho(A)(v) = \lambda(A) \cdot v\} \ne \{0\}$. By the lemma above, W is stable under $\mathfrak{g} = k \cdot X \oplus \mathfrak{a}$ (for any $X \in \mathfrak{g} - \mathfrak{a}$). Clearly, any non-zero eigen-vector for $\rho(X)$ in W is an eigen-vector for all of $\rho(\mathfrak{g})$. Hence the theorem.

The three corollaries below obviously follow from Lie's theorem.

6.66. Corollary. *Assume that $k = \bar{k}$. Then for any irreducible representation ρ of a solvable Lie algebra $\mathfrak{g}$, $d_\rho = 1$.*

6.67. Corollary. *Assume that $k = \bar{k}$ and $\mathfrak{g}$ is solvable. Then for any representation ρ of $\mathfrak{g}$, there are sub-representations $\{V_{\rho_i} | \ 1 \le i \le d_\rho\}$ of V_ρ with $V_{\rho_1} = \{0\}, V_{d_\rho} = V_\rho$, $V_{\rho_i} \subset V_{\rho_{(i+1)}}$ and $d_{\rho_i} = i$. Also $(\rho_{(i+1)}/\rho_i)(D^1(\mathfrak{g})) = \{0\}$. Equivalently, V_ρ admits an ordered basis with respect to which the matrices corresponding to all the endomorphisms in $\rho(\mathfrak{g})$ are upper triangular and $\rho(D^1(\mathfrak{g}))$ consists of nilpotent matrices.*

6.68. Corollary. *If $k = \bar{k}$, for any solvable Lie subalgebra $\mathfrak{g}$ of $M(n, k)$ there is a $g \in GL(n, k)$ such that $g \cdot (\mathfrak{g}) \cdot g^{-1}$ is contained in the Lie algebra of all upper triangular matrices in $M(n, k)$.*

6.69. Corollary. *If $\mathfrak{g}$ is a solvable Lie subalgebra of $M(n,k)$, $ld(\mathfrak{g}) \leq [(n+1)/2] + 1$ ($ld(\mathfrak{g})$ is the length of the derived series of $\mathfrak{g}$).*

6.70. Proof. There is no loss of generality in assuming that $k = \bar{k}$; and in that case $\mathfrak{g}$ can be assumed to be a subalgebra of the algebra $\mathfrak{t}$ of all upper triangular matrices. Thus $ld(\mathfrak{g}) \leq ld(\mathfrak{t})$. The corollary follows since $ld(\mathfrak{t}) = [(n+1)/2] + 1$ (exercise).

6.71. Proposition. *Let $\mathfrak{g}$ be solvable. Then for any representation ρ of $\mathfrak{g}$, $\rho(D^1(\mathfrak{g}))$ consists of nilpotent matrices and there are $\mathfrak{g}$-submodules $\{V_{\rho_i}|\ 1 \leq i \leq d_\rho\}$ with $V_{\rho_i} \subset V_{\rho_{(i+1)}}$, $V_{\rho_1} = \{0\}$ and $V_{d_\rho} = V_\rho$ such that $(\rho_{(i+1)}/\rho_i)(D^1(\mathfrak{g})) = \{0\}$ for $1 \leq i < d_\rho$.*

6.72. Proof. $\bar{\mathfrak{g}} = \mathfrak{g} \otimes_k \bar{k}$ is solvable and hence by the corollary above $\rho(D^1(\mathfrak{g}))$ consists entirely of nilpotent endomorphisms of $V_\rho \otimes \bar{k}$, hence of V_ρ. The last assertion now follows by combining Engel's Theorem and Dynkin's Lemma (6.62) (Any Lie algebra homomorphism of $\mathfrak{g}$ in k is zero on the ideal $D^1(\mathfrak{g})$).

6.73. Corollary. *If ρ is a completely reducible representation of a solvable Lie algebra $\mathfrak{g}$, $\rho(\mathfrak{g})$ is abelian and $\rho(D^1(\mathfrak{g})) = 0$.*

6.74. Corollary. *If $\mathfrak{g}$ is solvable, $D^1(\mathfrak{g})$ is a nilpotent ideal in $\mathfrak{g}$.*

6.75. Proof. Let $\mathfrak{c}$ be the centre of $\mathfrak{g}$. Then $\mathfrak{c}' = \mathfrak{c} \cap D^1(\mathfrak{g})$ is evidently central in $D^1(\mathfrak{g})$. Now $ad_\mathfrak{g}(D^1(\mathfrak{g}))(\simeq D^1(\mathfrak{g})/\mathfrak{c}')$ consists entirely of nilpotent matrices and is thus a nilpotent Lie algebra. Hence $D^1(\mathfrak{g})$ is nilpotent modulo a central ideal and is therefore itself nilpotent.

Yet another interesting consequence of Lie's theorem is the following.

6.76. Proposition. *Let ρ be a completely reducible representation of a Lie algebra $\mathfrak{g}$. Then $\rho([\mathfrak{g}, \mathfrak{r}(\mathfrak{g})]) = \rho([\mathfrak{g}, \mathfrak{g}] \cap \mathfrak{r}(\mathfrak{g})) = \{0\}$ ($\mathfrak{r}(\mathfrak{g})$ is the radical of $\mathfrak{g}$) (cf. 6.30).*

6.77. Proof. Set $\mathfrak{r} = \mathfrak{r}(\mathfrak{g})$. If ρ is a completely reducible representation of $\mathfrak{g}$, then $\bar{\rho}$ the extension of ρ to $\bar{\mathfrak{g}} = \mathfrak{g} \otimes_k \bar{k}$ remains completely reducible(see 6.7). We may hence assume $k = \bar{k}$. Clearly, we may also assume ρ to be irreducible. By Lie's theorem, there is a $0 \neq v_o \in V_\rho$ and a k-linear map $\lambda : \mathfrak{r} \to k$ such that $\rho(X)(v_o) = \lambda(X) \cdot v_o$ for all $X \in \mathfrak{r}$. By Dynkin's Lemma, the subspace $W = \{v \in V_\rho|\ \rho(X)(v) = \lambda(X) \cdot v, \forall X \in \mathfrak{r}\}$ is stable under $\rho(\mathfrak{g})$; $W \neq \{0\}$ as $v_o \in W$. As ρ is irreducible, $W = V_\rho$. Now if $X \in \mathfrak{r}$, $\text{Trace}(\rho(X)) = q \cdot \lambda(X)$ where $q = \dim_k V_\rho$; on the other hand if $X \in [\mathfrak{g}, \mathfrak{g}]$, $\text{Trace}(\rho(X)) = 0$. Thus $\rho(X) = 0$ if $X \in ([\mathfrak{g}, \mathfrak{g}] \cap \mathfrak{r})$ and $[\mathfrak{g}, \mathfrak{r}] \subset [\mathfrak{g}, \mathfrak{g}] \cap \mathfrak{r}$, proving the proposition.

6.78. Corollary. *Let ρ be a representation of a Lie algebra $\mathfrak{g}$. Then $\rho([\mathfrak{g}, \mathfrak{g}] \cap \mathfrak{r}(\mathfrak{g}))$ consists entirely of nilpotent matrices and is (hence) a nilpotent Lie algebra.*

Let $\{\rho_i | \ i \in I\}$ be the Jordan-Hölder series of ρ and $\tau = \coprod_{i \in I} \rho_i$. The proposition says that $\tau([\mathfrak{g}, \mathfrak{g}] \cap \mathfrak{r}(\mathfrak{g})) = \{0\}$. The corollary follows.

6.79. Corollary. $[\mathfrak{g}, \mathfrak{g}] \cap \mathfrak{r}(\mathfrak{g}) \subset \mathfrak{n}(\mathfrak{g})$ *where $\mathfrak{n}(\mathfrak{g})$ is the nil-radical of $\mathfrak{g}$ (see 6.44)).*

Proposition 6.74 combined with the previous corollary (applied to the adjoint representation) shows that $ad_{\mathfrak{g}}([\mathfrak{g}, \mathfrak{g}] \cap \mathfrak{r}(\mathfrak{g}))$ is a nilpotent Lie algebra. $[\mathfrak{g}, \mathfrak{g}] \cap \mathfrak{r}(\mathfrak{g}) \cap \mathfrak{c}(\mathfrak{g})$ is a central ideal in the algebra $[\mathfrak{g}, \mathfrak{g}] \cap \mathfrak{r}(\mathfrak{g})$ and the quotient by this ideal of $\mathfrak{g} \cap \mathfrak{r}(\mathfrak{g})$ is the nilpotent algebra $ad_{\mathfrak{g}}(\mathfrak{g} \cap \mathfrak{r}(\mathfrak{g}))$. Thus $([\mathfrak{g}, \mathfrak{g}] \cap \mathfrak{r}(\mathfrak{g}))$ is a nilpotent ideal in $\mathfrak{g}$ and hence contained in $\mathfrak{n}(\mathfrak{g})$.

6.80. Implications of Lie's Theorem for Lie Groups. Let k be a local field, G a Lie group over k and $\mathfrak{g}$ its Lie algebra. Let ρ be a representation of G. Suppose that $\mathfrak{g}$ is solvable. Let K be the field obtained by adjoining to k all the eigen-values of $\{\dot{\rho}(X) | \ X \in \mathfrak{g}\}$. This is a finite extension of k and is hence a local field (cf. 1.87). Now from Lie's theorem, we conclude that there is a linear map $\alpha : \mathfrak{g} \to K$ such that $V(\alpha) = \{v \in V_\rho \otimes_k K | \ \dot{\rho}(X) = \alpha(X) \cdot v\} \neq \{0\}$. We then have the following theorem as an immediate consequence of Lie's theorem.

6.81. Theorem. *Let ρ be an <u>admissible</u> representation of a Lie group G over the (local field) k (cf. 5.34). Assume that the Lie algebra $\mathfrak{g}$ of G is solvable. Let K be a finite extension of k containing all the eigen-values of $\dot{\rho}(X)$ for every $X \in \mathfrak{g}$ (cf. 1.95). Then there is a filtration $\{0\} = W_0 \subset W_1 \subset W_d \cdots \subset W_q = V_\rho \otimes_k K$ of $V_\rho \otimes_k K$ by G-stable K-vector subspaces W_i and analytic group morphism $\{\chi_i : G \to K^\times | \ 1 \leq i \leq q\}$ such that for $g \in G$ and $v \in W_i / W_{i-1}$, one has $\tau_i(g)(v) = \chi_i(g) \cdot v$ where τ_i is the representation (over K) of G on W_i / W_{i-1}.*

6.82. Corollary. *Let G, ρ, K and W be as in Theorem 6.81. Then there is an ordered basis B of W over K such that the matrix of $\rho(g)$ with respect to B is upper triangular for every $g \in G$. Further $\rho(G)$ is solvable and $\rho([G, G])$ consists entirely of unipotent transformations.*

6.83. Corollary. *If ρ is irreducible and admissible, $\rho(G)$ is abelian.*

6.84. Corollary. *If $k \simeq \mathbb{C}$ and G is connected and solvable, any irreducible representation ρ of G is of dimension 1.*

This is immediate from Schur's Lemma (1.53).

6.85. Corollary. *If $k = \mathbb{R}$, G is connected solvable and ρ is irreducible, then $\dim_k V_\rho \leq 2$. If $\dim_{\mathbb{R}} V = 1$, there is an analytic homomorphism $\chi : G \to \mathbb{R}^+$ such that $\rho(g)(v) = \chi(g) \cdot v$, $\forall\, g \in G, \forall v \in V$.*

6.86. Proof. Let $W = V_\rho \otimes_{\mathbb{R}} \mathbb{C}$ and E, an irreducible sub-representation (over $\mathbb{C}$) of W. Let σ denote complex conjugation in $\mathbb{C}$ and $\sigma_W = 1_{V_\rho} \otimes \sigma$. Then as $E + \sigma_W(E)$ is stable under σ_W, it is necessarily of the form $V' \otimes_{\mathbb{R}} \mathbb{C}$ for some $\mathbb{R}$-vector subspace V' of V. Since $E + \sigma_W(E)$ is G-stable so is V'. It follows that $V' = V$. Evidently, $\dim_{\mathbb{R}} V' \leq 2$. Thus $\dim_\rho V \leq 2$. The last assertion follows from the fact that $\rho(G)$ is connected and is hence contained in $\mathbb{R}^+$.

The following theorem is an easy consequence of the definitions.

6.87. Theorem. *Let G be a Lie group and ρ a representation of G. Let G_o be the group generated by $\rho^{-1}(\exp_{GL(V_\rho)})(\Omega_{\mathfrak{gl}(V_\rho)})$. Then G_o is an open subgroup of G and $\rho|_{G_o}$ is admissible.*

The next result holds the key to an in-depth study of the structure of Lie algebras.

6.88. Theorem (Cartan's Criterion for Solvability). *Let $\mathfrak{g}$ be a Lie subalgebra of $\mathfrak{gl}(V)$. If $\operatorname{trace}(X \cdot Y) = 0$, $\forall X, Y \in \mathfrak{g}$, $\mathfrak{g}$ is solvable.*

6.89. Proof. We assume, as we may, that k is algebraically closed. Let $\mathcal{L}$ be the set of all Lie subalgebras of $\mathfrak{gl}(V)$ which are not solvable and on which $B_V \equiv 0$. We need to show that $\mathcal{L} = \phi$. If not, let $\mathfrak{g} \in \mathcal{N}$ be of minimal dimension. As $\mathfrak{g}$ is not solvable, $[\mathfrak{g}, \mathfrak{g}]$ is not solvable; and the minimality of $\dim \mathfrak{g}$ implies that $\mathfrak{g} = [\mathfrak{g}, \mathfrak{g}]$. Let $\mathfrak{b}$ be a maximal solvable subalgebra of $\mathfrak{g}$. Let τ denote the quotient of the representation $ad_{\mathfrak{g}}|_{\mathfrak{b}}$ (of $\mathfrak{b}$ on $\mathfrak{g}$) by the $ad_{\mathfrak{b}}$. Then since $\mathfrak{b}$ is solvable, there is a linear form α on $\mathfrak{b}$ and a vector $E \in \mathfrak{g} - \mathfrak{b}$ such that
$$[B, E] = \alpha(B) \cdot E + \Phi(B), \quad \forall B \in \mathfrak{b}$$
where Φ is a linear map of $\mathfrak{b}$ into itself (Lie's theorem). Hence $\mathfrak{b}' = \mathfrak{b} + k \cdot E$ is a subalgebra of $\mathfrak{g}$ containing $\mathfrak{b}$ properly. Since $\mathfrak{b}$ is a maximal solvable subalgebra of $\mathfrak{g}$, $\mathfrak{b}'$ is not solvable; hence $\alpha \neq 0$. By the minimality of $\dim \mathfrak{g}$, $\mathfrak{b}' = \mathfrak{g}$. Thus $\mathfrak{b}$ has co-dimension 1 in $\mathfrak{g}$. Fix an element $H \in \mathfrak{b}$ such that $\alpha(H) = 2$.

Now suppose that ρ is any irreducible representation of $\mathfrak{g}$ on a vector space $W(= V_\rho)$. By Lie's theorem, there is a linear form λ on $\mathfrak{b}$ and a vector $w_0 \in W$ such that $\rho(B)(w_0) = \lambda(B) \cdot w_0$, $\forall B \in \mathfrak{b}$. Let n be the maximal integer ≥ 0 such that $S = \{w_q = \rho(E)^q(w_0) \mid 0 \leq q \leq n\}$ are linearly independent and W' the linear span of S. We will now prove by induction on q that for $B \in \mathfrak{b}$ and $0 \leq q \leq n$,
$$\rho(B)w_q = (\lambda + q \cdot \alpha)(B) \cdot w_q + f_q(B) \tag{$*$}$$

where $f_q(B) \in W_{q-1}(= $ Linear span of $\{w_i| \ 0 \le i \le (q-1)\})$. Assume that $(*)$ holds for all $q \le r$; then for $B \in \mathfrak{b}$,

$$\rho(B)(w_{(r+1)}) = \rho(B)\rho(E)(w_r) = \rho(E)\rho(B)w_r + \rho(\alpha(B) \cdot E + \Phi(B))(w_r)$$

$$= \rho(E)((\lambda + (r+1) \cdot \alpha)(B) \cdot w_r + f_r(B)) + \Phi(B)(w_r)$$

and $\Phi(B)(w_r) = (\lambda + r \cdot \alpha)(\Phi(B))(w_r) + f_r(\Phi(B))$ (induction hypothesis). Set $f_{(r+1)}(B) = f_r(B) + f_r(\Phi(B))$, Then $(*)$ holds for $q = (r+1)$. This proof takes care of the start of the induction as well. Thus W' is stable under $\rho(E)$ and $\rho(\mathfrak{b})$, hence under $\rho(\mathfrak{g})$. Thus $W' = V_\rho$. From $(*)$, we see that if $B \in \mathfrak{b}$, the eigen-values of $\rho(B)$ are $\{\lambda(B) + q \cdot \alpha(B)| \ 0 \le q \le n\}$, all of multiplicity 1. Since $B \in [\mathfrak{g}, \mathfrak{g}] = \mathfrak{g}$, we have

$$0 = \mathrm{trace}\rho(B) = (n+1) \cdot \lambda(B) + (n \cdot (n+1)/2) \cdot \alpha(B),$$

leading to $\lambda(B) = -n \cdot \alpha(B)/2$. Thus for *any* irreducible representation ρ of $\mathfrak{g}$ the eigen-values of $\rho(H)$ are integers. The eigen-values of H acting on V are the same as those of H acting on the direct sum of the irreducible representations ρ_i occurring in the Jordan-Hölder series of V and so the eigen-values of H are all in $\mathbb{Z}$. If all these were 0, so would be all the eigen-values of $ad_{\mathfrak{gl}(V)}(H)$ on $\mathfrak{gl}(V)$, a contradiction since 2 is an eigen-value of $ad_{\mathfrak{gl}(V)}(H)$. Thus $B_V(H, H) > 0$, a contradiction as $B_V \equiv 0$. Hence the theorem.

6.90. Remarks. (**1**) A Lie algebra $\mathfrak{g}$ is solvable if its Killing form is zero. By the Cartan criterion $ad(\mathfrak{g}) \subset \mathfrak{gl}(\mathfrak{g})$ is solvable. The kernel of ad is abelian, as it is the centre of $\mathfrak{g}$. Hence $\mathfrak{g}$ is solvable.

(**2**) If $\mathfrak{g}$ is a Lie subalgebra of $\mathfrak{gl}(V)$ consisting entirely of nilpotent matrices, then the Trace form on $\mathfrak{gl}(V)$ is zero on $\mathfrak{g}$ (cf. 6.52). However, the trace form need not be identically zero on (even) an abelian subalgebra. The simplest example is the 1-dimensional algebra $\mathfrak{gl}(1, k)$ – in fact, in this case the trace form is non-degenerate.

(**3**) On the other hand there are non-nilpotent solvable Lie subalgebras of $\mathfrak{sl}(3, k)$ with k a field containing $\sqrt{2}$ and $\sqrt{-1}$ on which the trace form is identically zero. Here is an example: Let $X \in \mathfrak{gl}(3, \mathbb{C})$ be the diagonal matrix with $X_{11} = \sqrt{2}, X_{22} = X_{33} = \sqrt{-1}$. Let $\mathfrak{u}$ be the Lie algebra of upper triangular nilpotent matrices. Then $[X, \mathfrak{u}] \subset \mathfrak{u}$, so that $\mathfrak{g} = (k \cdot X) \oplus \mathfrak{u}$ is a Lie subalgebra of $\mathfrak{gl}(3, k)$. It is easily checked that the trace form of $\mathfrak{gl}(V)$ is identically zero on $\mathfrak{g}$; and $\mathfrak{g}$ is a solvable but not nilpotent, Lie algebra.

Cartan's criterion has the following implication for Lie groups.

6.91. Theorem. *Let G be a Lie group over a local field k and $\rho : G \to GL(V)$ an admissible representation. Suppose that the Trace bilinear form $(X, Y) \rightsquigarrow \mathrm{Trace}(\rho(X)) \cdot (\rho(Y)), X, Y \in \mathfrak{g}$ is zero. Then $\rho(G)$ is solvable.*

6.92. Semi-Simple Lie Algebras. A Lie algebra $\mathfrak{g}$ is *semi-simple*, if its radical $\mathfrak{r}(\mathfrak{g}) = 0$. The last non-zero term of the derived series of a solvable ideal $\mathfrak{s}$ in a Lie algebra $\mathfrak{g}$ is an abelian ideal in $\mathfrak{g}$; thus $\mathfrak{g}$ is semi-simple iff it admits no non-zero abelian ideals. If $\mathfrak{g}$ is any Lie algebra, the Lie algebra $\mathfrak{h} = \mathfrak{g}/\mathfrak{r}(\mathfrak{g})$ is semi-simple. This is because, the inverse image of $\mathfrak{r}(\mathfrak{h})$ in $\mathfrak{g}$ is solvable and hence contained in $\mathfrak{r}(\mathfrak{g})$. A Lie algebra $\mathfrak{g}$ is *reductive*, if its radical $\mathfrak{r}(\mathfrak{g}) = \mathfrak{c}(\mathfrak{g})$, the centre of $\mathfrak{g}$.

A Lie algebra $\mathfrak{g}$ is *simple* if it is not abelian and has no proper ideal $\neq \{0\}$. A simple Lie algebra is semi-simple. More generally, a direct product of any finite set of simple Lie algebra is semi-simple. A Lie group over a local field is *semi-simple* if its Lie algebra is semi-simple.

The following characterization of a semi-simple Lie algebra, is the crucial first step in the study of the structure of semi-simple Lie algebras which, however, we will not go into in this book.

6.93. Theorem. *A Lie algebra $\mathfrak{g}$ is semi-simple, if and only if the Killing form $B_{\mathfrak{g}}$ is non-degenerate.*

6.94. Proof. Let $\mathfrak{a} = \{X \in \mathfrak{g} \mid B_{\mathfrak{g}}(X, Y) = 0, \ \forall \ Y \in \mathfrak{g}\}$. Then $\mathfrak{a}$ is an ideal in $\mathfrak{g}$ and $ad_{\mathfrak{g}}(\mathfrak{a})$ is solvable (Theorem 6.88). The kernel $\mathfrak{c}$ of $ad_{\mathfrak{g}}$ being the centre of $\mathfrak{g}$, $\mathfrak{c} \cap \mathfrak{a}$ is solvable. Thus $\mathfrak{a}$ is solvable. Hence if $\mathfrak{g}$ is semi-simple $\mathfrak{a} = \{0\}$, i.e., the Killing form is non-degenerate. Conversely, suppose that $B_{\mathfrak{g}}$ is non-degenerate and $\mathfrak{a}$ is an abelian ideal; then one has for $A \in \mathfrak{a}$ and $X \in \mathfrak{g}$, $(ad_{\mathfrak{g}}(A) \cdot ad_{\mathfrak{g}}(X))(\mathfrak{g}) \subset \mathfrak{a}$ and $(ad_{\mathfrak{g}}(A) \cdot ad_{\mathfrak{g}}(X))(\mathfrak{a}) = 0$ so that $(ad_{\mathfrak{g}}(A) \cdot ad_{\mathfrak{g}}(X))^2 = 0$. Thus if $A \in \mathfrak{a}$, $B_{\mathfrak{g}}(A, X) = 0$ for all $X \in \mathfrak{g}$; as B is non-degenerate, $A = 0$. Hence $\mathfrak{a} = \{0\}$, i.e., there is no non-zero abelian ideal in $\mathfrak{g}$. Hence the theorem.

6.95. Corollary. *Let $\mathfrak{g}$ be a semi-simple Lie algebra. Then the adjoint representation of $\mathfrak{g}$ is completely reducible. The irreducible $\mathfrak{g}$-submodules of $\mathfrak{g}$ are precisely the minimal ideals. Every minimal ideal is simple as a Lie algebra. Let $\mathcal{M}$ be the set of minimal ideals in $\mathfrak{g}$. Then $\mathfrak{g}$ is the direct product $\prod_{\mathfrak{a} \in \mathcal{M}} \mathfrak{a}$ of all its minimal ideals.*

6.96. Proof. Let $\mathfrak{b}$ be any non-zero ideal in $\mathfrak{g}$ and $\mathfrak{r}(\mathfrak{b})$ its radical. Since $\mathfrak{r}(\mathfrak{b})$ is a characteristic ideal in $\mathfrak{b}$, it is an ideal in $\mathfrak{g}$; and as $\mathfrak{g}$ is semi-simple, $\mathfrak{r}(\mathfrak{b}) = \{0\}$. Let $\mathfrak{g}'$ be the sum of all the minimal ideals in $\mathfrak{g}$. The killing form of $\mathfrak{g}$ restricted to $\mathfrak{g}'$ is the Killing form of $\mathfrak{g}'$. Since $\mathfrak{g}'$ is semi-simple, $B_{\mathfrak{g}}$ restricted to $\mathfrak{g}'$ is non-degenerate. It follows that $\mathfrak{g} = \mathfrak{g}' \oplus \mathfrak{g}'^{\perp}$ where $\mathfrak{g}'^{\perp} = \{X \in \mathfrak{g} \mid B_{\mathfrak{g}}(X, Y) = 0, \ \forall Y \in \mathfrak{g}'\}$. If $\mathfrak{g}'^{\perp} \neq \{0\}$, it will contain a non-zero minimal ideal, a contradiction. Thus $\mathfrak{g}' = \mathfrak{g}$. That $\mathfrak{g}$ is a direct product of all its minimal ideals, now follows from the fact that for any two distinct non-zero minimal ideals $\mathfrak{a}$, $\mathfrak{b}$ in $\mathfrak{g}$, $[\mathfrak{a}, \mathfrak{b}] \subset (\mathfrak{a} \cap \mathfrak{b}) = \{0\}$.

6.97. Corollary. *Let V be a vector space and $\mathfrak{g} \subset \mathfrak{gl}(V)$ a Lie subalgebra such that $B_V|_\mathfrak{g}$ is non-degenerate. Then $\mathfrak{g}$ is reductive and hence a direct product of its centre $\mathfrak{c}(\mathfrak{g})$ and the commutator ideal $[\mathfrak{g}, \mathfrak{g}]$.*

6.98. Proof. We assume, as we may, that k is algebraically closed. Let ρ be the representation of $\mathfrak{g}$ on V given by the inclusion of $\mathfrak{g}$ in $\mathfrak{gl}(V)$. Let $\{\rho_i | i \in I\}$ be the Jordan-Hölder series of ρ and $\bar{\rho} = \coprod_{i \in I} \rho_i$. Then $B_V|_\mathfrak{g} = B_\rho = B_{\bar{\rho}}$. Hence the kernel of $\bar{\rho}$ is trivial. As the ρ_i are irreducible, Lie's theorem combined with Dynkin's lemma shows that $\rho_i(\mathfrak{r}(\mathfrak{g}))$ is central in $\rho_i(\mathfrak{g})$ for all $i \in I$. Since $\bar{\rho}$ is faithful, it follows that $\mathfrak{r}(\mathfrak{g})$ is central in $\mathfrak{g}$. Now $\mathfrak{h} = \mathfrak{g}/\mathfrak{r}(\mathfrak{g})$ being semi-simple, $[\mathfrak{h}, \mathfrak{h}] = \mathfrak{h}$ (6.95); hence $[\mathfrak{g}, \mathfrak{g}]$ maps onto $\mathfrak{h}$. Now for every $i \in I$, $\rho_i([\mathfrak{g}, \mathfrak{g}]) \subset \mathfrak{sl}(V_{\rho_i})$ while $\rho_i(\mathfrak{r}(\mathfrak{g})) \subset$ scalars in $\mathfrak{gl}(V_{\rho_i})$. It follows that $\bar{\rho}[\mathfrak{g}, \mathfrak{g}] \cap \bar{\rho}(\mathfrak{r}(\mathfrak{g})) = \{0\}$ and since $\bar{\rho}$ is faithful, $[\mathfrak{g}, \mathfrak{g}] \cap \mathfrak{r}(\mathfrak{g}) = \{0\}$. It follows that $[\mathfrak{g}, \mathfrak{g}] \simeq \mathfrak{h}$ is semi-simple and $\mathfrak{g} \simeq \mathfrak{h} \times \mathfrak{r}(\mathfrak{g})$, proving the corollary.

6.99. Corollary. *The radical $\mathfrak{r}(\mathfrak{g})$ and the nil-radical $\mathfrak{n}(\mathfrak{g})$ of a Lie algebra $\mathfrak{g}$ are characteristic ideals.*

6.100. Proof. We argue by induction on $\dim_k \mathfrak{g}$ to prove that $\mathfrak{r}(\mathfrak{g})$ is a characteristic ideal. Let $\mathfrak{a}$ be the orthogonal complement of $\mathfrak{g}$ with respect to the Killing form of $\mathfrak{g}$: $\mathfrak{a} = \{X \in \mathfrak{g} | B_\mathfrak{g}(X, Y) = 0, \forall Y \in \mathfrak{g}\}$. If $\mathfrak{a} = \{0\}$, $B_\mathfrak{g}$ is non-degenerate, $\mathfrak{r}(\mathfrak{g}) = \mathfrak{n}(\mathfrak{g}) = \{0\}$ and the corollary is trivially true. Assume then $\mathfrak{a} \neq \{0\}$. Now $\mathfrak{a}$ is a characteristic ideal $\mathfrak{g}$ (see 6.21). By Theorem 6.88, $ad(\mathfrak{a})$ is solvable. The kernel of ad is central in $\mathfrak{g}$. Hence $\mathfrak{a}$ is solvable. Let $\mathfrak{h} = \mathfrak{g}/\mathfrak{a}$; then $\dim_k \mathfrak{h} < \dim_k \mathfrak{g}$ and by induction hypothesis, $\mathfrak{r}(\mathfrak{h})$ is a characteristic ideal in $\mathfrak{h}$. Now if D is a derivation of $\mathfrak{g}$, it stabilizes $\mathfrak{a}$ and hence defines a derivation $\bar{D}$ of $\mathfrak{h}$ and hence $\bar{D}(\mathfrak{r}(\mathfrak{h})) \subset \mathfrak{r}(\mathfrak{h})$. It follows that $D(\pi^{-1}(\mathfrak{r}(\mathfrak{h}))) \subset \pi^{-1}(\mathfrak{r}(\mathfrak{h}))$; and $\pi^{-1}(\mathfrak{r}(\mathfrak{h})) = \mathfrak{r}(\mathfrak{g})$. Thus $\mathfrak{r}(\mathfrak{g})$ is a characteristic ideal. It follows that $[\mathfrak{g}, \mathfrak{r}(\mathfrak{g})]$ is a characteristic ideal.

If $[\mathfrak{g}, \mathfrak{r}(\mathfrak{g})] = 0$, $\mathfrak{r}(\mathfrak{g})$ is central in $\mathfrak{g}$ and hence abelian, therefore nilpotent. In this case, $\mathfrak{n}(\mathfrak{g}) = \mathfrak{r}(\mathfrak{g})$ and is hence a characteristic ideal. To prove that $\mathfrak{n}(\mathfrak{g})$ is characteristic in general, we argue again by induction on $\dim_k \mathfrak{g}$. Assume that this holds for all $\mathfrak{g}$ of dimension $\leq n$ and suppose that $\mathfrak{g}$ is of dimension $n + 1$. We saw that if $[\mathfrak{g}, \mathfrak{r}(\mathfrak{g})] = \{0\}$, $\mathfrak{n}(\mathfrak{g}) = \mathfrak{r}(\mathfrak{g})$ is a characteristic ideal. Assume that $\mathfrak{s} = [\mathfrak{g}, \mathfrak{r}(\mathfrak{g})] \neq \{0\}$. Let $\mathfrak{b} = DC^q(\mathfrak{s})$ be the last non-zero ideal in the descending central series of $\mathfrak{s}$ and set $\mathfrak{b}' = DC^{q-1}(\mathfrak{s})$; $\mathfrak{b}$ and $\mathfrak{b}'$ are characteristic ideals in $\mathfrak{g}$. Now the map $(X, Y) \rightsquigarrow [X, Y]$ of $\mathfrak{s} \times \mathfrak{b}'$ in $\mathfrak{b}$ factors through $(\mathfrak{s}/\mathfrak{b}) \times (\mathfrak{b}'/\mathfrak{b})$ and is compatible with the action of $\mathfrak{g}$ (via $\mathfrak{g}/\mathfrak{b}$) yielding a $\mathfrak{g}/\mathfrak{b}$-module homomorphism of $\mathfrak{s}/\mathfrak{b} \otimes \mathfrak{b}'/\mathfrak{b}$ in $\mathfrak{b}$. It follows that if $X \in \mathfrak{r}(\mathfrak{g})$

is such that $ad\bar{X}$ ($\bar{X}$ is the image of $X \in \mathfrak{g}/\mathfrak{b}$) acts nilpotently on $\mathfrak{g}/\mathfrak{b}$, adX acts nilpotently on $\mathfrak{b}$ as well, hence on $\mathfrak{g}$ as well. It follows that $X \in \mathfrak{n}(\mathfrak{g})$ if and only if $\bar{X} \in \mathfrak{n}(\mathfrak{g}/\mathfrak{b})$. By the induction hypothesis ($\dim_k \mathfrak{g}/\mathfrak{b} \leq n$), $\mathfrak{n}(\mathfrak{g}/\mathfrak{b})$ is a characteristic ideal. As $\mathfrak{b}$ is a characteristic ideal, $\mathfrak{n}(\mathfrak{g})$, the inverse image of $\mathfrak{n}(\mathfrak{g}/\mathfrak{b})$ in $\mathfrak{g}$ is a characteristic ideal.

7. Lie Algebras: Structure Theory

In this chapter we prove some basic results about semi-simple Lie algebras. The first of these is that every finite dimensional representation of a semi-simple Lie algebra is completely reducible. This was first proved by Hermann Weyl who deduced it from the complete reducibility of representations of compact real Lie groups. J H C Whitehead came up with a purely algebraic proof which also yielded the fact that every finite dimensional Lie algebra is the semi-direct product of a semi-simple subalgebra and the radical of the Lie algebra. The proof given here is Bourbaki's adaptation of Whitehead's proof. The chapter (and the book) ends with a proof of Ado's theorem which says that every finite dimensional Lie algebra admits a faithful finite dimensional representation.

We continue with the notations of Chapter 6. We now prove a key result about representations of a semi-simple Lie algebra $\mathfrak{g}$ that will enable us to deduce important properties of Lie algebras in general.

7.1. Theorem. *Let $\mathfrak{g}$ be a semi-simple Lie algebra and $\rho : \mathfrak{g} \to \mathfrak{gl}(V)$ be a representation of $\mathfrak{g}$. Then $\mathfrak{a} = \{X \in \mathfrak{g} | \ B_\rho(X, Y) = \text{trace}(\rho(X) \cdot \rho(Y)) = 0, \ \forall Y \in \mathfrak{g}\}$ is an ideal in $\mathfrak{g}$. If $\mathfrak{b}$ is the (unique) ideal in $\mathfrak{g}$ such that $\mathfrak{a} \oplus \mathfrak{b} = \mathfrak{g}$, then B_ρ restricted to $\mathfrak{b}$ is non-degenerate. Moreover, if $\{X_i | 1 \leq i \leq n\}$ and $\{Y_i | 1 \leq i \leq n\}$ are ordered bases of $\mathfrak{b}$ such that $B_\rho(X_i, Y_j) = \delta_{i,j}$ for $1 \leq i, j \leq n$, then if we set $C(\rho) = \sum_{1 \leq i \leq n} \rho(X_i) \cdot \rho(Y_i)$, we have for all $Z \in \mathfrak{g}$, $\rho(Z) \cdot C(\rho) = C(\rho) \cdot \rho(Z)$.*

7.2. Proof. For $X, Y, Z \in \mathfrak{g}$, $B_\rho([X, Y], Z) + B_\rho(Y, [X, Z]) = 0$. Thus if $Y \in \mathfrak{a}$, $B_\rho([X, Y], Z) = -B_\rho(Y, [X, Z]) = 0$ for all $X, Z \in \mathfrak{g}$. Hence $\mathfrak{a}$ is an ideal. In view of the complete reducibility of the adjoint representation (6.95), $\mathfrak{g} = \mathfrak{a} \oplus \mathfrak{b}$ for a unique ideal $\mathfrak{b}$ in $\mathfrak{g}$. Since $\mathfrak{a}$ is the kernel of B_ρ, B_ρ restricted to $\mathfrak{b}$ is non-degenerate. As $[\mathfrak{a}, \mathfrak{b}] \subset \mathfrak{a} \cap \mathfrak{b} = \{0\}$, $\mathfrak{a}$ and $\mathfrak{b}$ commute. Further since B_ρ is zero on $\mathfrak{a}$, $\rho(\mathfrak{a})$ is solvable (6.88) and hence zero. Next, for $1 \leq i \leq n$, we have for $Z \in \mathfrak{g}$,

$$[\rho(Z), \rho(X_i) \cdot \rho(Y_i)] = [\rho(Z), \rho(X_i)] \cdot \rho(Y_i) + \rho(X_i) \cdot [\rho(Z), \rho(Y_i)].$$

Now

$$[\rho(Z), \rho(X_i)] \cdot \rho(Y_i) = \rho(\sum_{j=1}^{n} B_\rho([Z, X_i], Y_j) \cdot X_j) \cdot \rho(Y_i)$$

while

$$\rho(X_i) \cdot [\rho(Z), \rho(Y_i)]) = \rho(X_i) \cdot (\sum_{j=1}^{n} B_\rho([Z, Y_i], X_j) \cdot Y_j).$$

As $B_\rho([Z, X_i], Y_j) = -B_\rho(X_i, [Z, Y_j])$, we conclude that $C(\rho) \cdot \rho(Z) = \rho(Z) \cdot C(\rho)$.

7.3. Corollary. *For a representation ρ of a semi-simple Lie algebra $\mathfrak{g}$, $C(\rho) = 0$ if and only if $\rho(\mathfrak{g}) = \{0\}$.*

7.4. Proof. If $\rho(\mathfrak{g}) \neq \{0\}$, the trace bilinear form $(X, Y) \rightsquigarrow \mathrm{Trace}(X \cdot Y)$ on $\mathrm{End}_k V_\rho$ restricted to $\rho(\mathfrak{g})$ is non-degenerate: otherwise, the kernel of this bilinear form on $\rho(\mathfrak{g})$ would be a solvable ideal in the semi-simple Lie algebra $\rho(\mathfrak{g})$. It follows from the definition of $C(\rho)$ that $\mathrm{trace} C(\rho) = \dim_k V_\rho$ and thus $C(\rho)$ is non-zero.

7.5. Theorem. *Every representation ρ of a semi-simple Lie algebra $\mathfrak{g}$ is completely reducible. Every representation of a semi-simple Lie group* (cf. 6.92) *of exponential type over a local field is completely reducible.*

7.6. Proof. The statement about Lie groups is immediate from the definitions and the complete reducibility of representations of semi-simple Lie algebras, which we proceed to prove now. We assume, as we may, that $k = \bar{k}$. We need to prove the following:
Let ρ be a representation of $\mathfrak{g}$ and $W \subset V_\rho$ a sub-representation of ρ (to be denoted ρ_1 in the sequel: $W = V_{\rho_1}$). Let ρ_2 be the quotient of ρ by ρ_1: $V_{\rho_2} = V_\rho/W$. Then there is a $\mathfrak{g}$-module homomorphism $q : V_{\rho_2} \to V_\rho$ such that $\pi {\circ} q$ is the Identity endomorphism $1_{V_{\rho_2}}$ of V_{ρ_2} where $\pi : V_\rho \to V_{\rho_2}$ is the natural map.
The exact sequence

$$\{0\} \to V_{\rho_1} \to V_\rho \to V_{\rho_2} \to \{0\}$$

of $\mathfrak{g}$-modules yields the following exact sequence (of $\mathfrak{g}$-modules):

$$\{0\} \to \mathrm{Hom}_k(V_{\rho_2}, V_{\rho_1}) \to \mathrm{Hom}_k(V_{\rho_2}, V_\rho) \xrightarrow{\pi^*} \mathrm{Hom}(V_{\rho_2}, V_{\rho_2}) \to \{0\}.$$

where π^* is defined by setting for $u \in \mathrm{Hom}_k(V_{\rho_2}, , V_\rho)$, $\pi^* u = \pi {\circ} u$. Let $E = \pi^{-1}(k \cdot 1_{V_{\rho_2}})$ so that we have the exact sequence of $\mathfrak{g}$-modules:

$$\{0\} \to \mathrm{Hom}_k(V_{\rho_2}, V_{\rho_1}) \to E \xrightarrow{\pi^*} k \cdot 1_{V_{\rho_2}} \to \{0\}.$$

The splitting of this sequence of $\mathfrak{g}$-modules means that there is a $\mathfrak{g}$ invariant element q in $\mathrm{Hom}_k(V_{\rho_2}, V_\rho)$ such that $\pi {\circ} q = 1_{V_{\rho_2}}$. In other words, it suffices to show that any exact sequence

$$\{0\} \to V_{\rho_1} \to V_\rho \to V_{\rho_2} \to \{0\} \qquad (*)$$

of $\mathfrak{g}$-modules with ρ_2 the trivial 1-dimensional representation, splits.

We will prove this by induction on the length of V_{ρ_1} as a $\mathfrak{g}$-module. At the start of the induction, when the length is 1, V_{ρ_1} is irreducible. If ρ_1 is trivial, since ρ_2 is also trivial, $\rho(\mathfrak{g})$ would consist entirely of nilpotent upper triangular matrices for a suitable basis of $V(\rho)$ and hence would be solvable. It follows that $\rho(\mathfrak{g}) = \{0\}$. Hence if $v \in V_\rho - V_{\rho_1}$, $k \cdot v$ is a sub-representation of ρ supplementary to V_{ρ_1}.

If V_{ρ_1} is non-trivial, $B_\rho = B_{\rho_1}$ is non-zero and the endomorphism $C(\rho_1)$ which is the restriction of $C(\rho)$ to V_{ρ_1} is non-zero. Note that $\text{trace}(C(\rho_1)) = n$ ($C(\rho)$ is defined in (7.3)). Now $C(\rho)$ is a $\mathfrak{g}$-module endomorphism whose restriction to V_{ρ_1} equals $C(\rho_1)$; and since V_{ρ_1} is irreducible (and $k = \bar{k}$), it is a scalar (Schur's Lemma (1.53)). Since $C(\rho)$ acts as zero on V_{ρ_2} (which is of dimension 1), 0 is an eigen-value of $C(\rho)$ of multiplicity 1. It follows that $V_\rho = V_\rho(0) \oplus V_{\rho_1}(= C(\rho)(V_\rho))$ where $V_\rho(0)$ is the kernel of $C(\rho)$ and is therefore stable under $\rho(\mathfrak{g})$. This takes care of the start of the induction.

Assume then that we have shown that the sequence $(*)$ is split when V_{ρ_1} has length $< l(\geq 2)$. Suppose then that ρ_1 has length l. Let r_1' be a non-zero irreducible sub-representation of r_1. Passing to the quotient by $V_{\rho_1'}$ we obtain from $(*)$, the exact sequence

$$\{0\} \to V_{\rho_1}/V_{\rho_1'} \to V_\rho/V_{\rho_1'} \to V_{\rho_2} \to \{0\}.$$

Now the length of $V_{\rho_1}/V_{\rho_1'} = l - 1$. By the induction hypothesis, we can find a $\mathfrak{g}$-sub-representation ρ_3 of $V_\rho/V_{\rho_1'}$ such that $V_\rho/V_{\rho_1'} = V_{\rho_1}/V_{\rho_1'} \oplus V_{\rho_3}$. Let V_{ρ_4} be the inverse image of (the trivial) representation V_{ρ_3} in V_ρ. We then have the exact sequence of $\mathfrak{g}$-modules:

$$\{0\} \to V_{\rho_1'} \to V_{\rho_4} \to V_{\rho_3} \to \{0\}$$

which splits (start of the induction). This proves the theorem.

7.7. Corollary. *Every derivation (see 6.11) D of a semi-simple Lie algebra $\mathfrak{g}$ is an inner derivation.*

7.8. Proof. Let $\mathfrak{g}'$ be the semi-direct product $(k \cdot D) \propto \mathfrak{g}$. Then $ad_{\mathfrak{g}'}|_\mathfrak{g}$ being completely reducible, $\mathfrak{g}' = \mathfrak{g} \oplus L$ where L is a $\mathfrak{g}$-submodule with $\dim_k L = 1$. Since $\mathfrak{g} = [\mathfrak{g}, \mathfrak{g}]$, L is a trivial $\mathfrak{g}$-module and thus L is central in $\mathfrak{g}'$. Now $D = D' + X$ with $D' \in L$ and $X \in \mathfrak{g}$. It follows that for $Y \in \mathfrak{g}$, $D(Y) = [(D'+X), Y]$ (in $\mathfrak{g}'$) $= [X, Y]$ – in other words, $D = adX$. Hence the corollary.

7.9. Corollary. *Let $\mathfrak{g}$ be a semi-simple Lie algebra and $X \in \mathfrak{g}$. Then there exist $X_s, X_n \in \mathfrak{g}$ such that $X = X_s + X_n$, $[X_s, X_n] = 0$ and, for every representation ρ of $\mathfrak{g}$, $\rho(X_s) = \rho(X)_s$, the semi-simple part of $\rho(X)$ and $\rho(X_n) = \rho(X)_n$, the nilpotent part of $\rho(X)$.*

7.10. Proof. We assume, as we may, that k is algebraically closed. Consider now the adjoint representation $ad_\mathfrak{g}$. Then $ad_\mathfrak{g}(X)$ being a derivation of $\mathfrak{g}$, so is $ad_\mathfrak{g}(X)_s$, the semi-simple part of $ad_\mathfrak{g}(X)$ (Proposition 6.19). By 7.7, there is a $X_s \in \mathfrak{g}$ such that $ad_\mathfrak{g}(X)_s = ad_\mathfrak{g}(X_s)$; and $ad_\mathfrak{g}(X)_n = ad_\mathfrak{g}(X) - ad_\mathfrak{g}(X)_s = ad_\mathfrak{g}(X) - ad_\mathfrak{g}(X_s) = ad_\mathfrak{g}(X - X_s)$. Set $X_n = X - X_s$. From the fact that $ad_\mathfrak{g}$ is faithful, we see that X_s and X_n are uniquely determined.

Now any representation ρ is a direct sum $\coprod_{i \in I} \rho_i$ with I finite and ρ_i irreducible. Then one has for $X \in \mathfrak{g}$, $\rho(X) = \coprod_{i \in I} \rho_i(X)$ and hence $\rho(X)_s = \coprod_{i \in I} \rho_i(X)_s$. It follows that it suffices to prove that for any *irreducible* ρ, $\rho(X)_s = \rho(X_s)$. When ρ is trivial, there is nothing to prove. Assume that ρ is non-trivial. Replacing $\mathfrak{g}$ by $\rho(\mathfrak{g})$, we assume that ρ is faithful. Now $X \in \mathfrak{g}$, $ad_{\mathfrak{gl}(V_\rho)}(\rho(X))$ stabilizes $\rho(\mathfrak{g}) \simeq \mathfrak{g}$ and its restriction to $\rho(\mathfrak{g})$ is a derivation of $\mathfrak{g}$. If $\rho(X)_s$ is the semi-simple part of $\rho(X)$, $ad_{\mathfrak{gl}(V_\rho)}(\rho(X))_s$ is the semi-simple part $ad_{\mathfrak{gl}(V_\rho)}(\rho(X))$. It follows that $ad_{\mathfrak{gl}(V_\rho)}(\rho(X)_s)$ stabilizes $\rho(\mathfrak{g})$ and is a derivation. This means that there is a $X_s' \in \mathfrak{g}$ such that $ad_{\mathfrak{gl}(V_\rho)}(\rho(X))_s$ restricted to $\mathfrak{g}$ equals $ad_\mathfrak{g}(X_s')$. Hence $ad_{\mathfrak{gl}(V_\rho)}(\rho(X_s - X_s'))$ is zero on all of $\mathfrak{g}$. In other words, $\rho(X_s - X_s')$ centralizes $\rho(\mathfrak{g})$. By Schur's Lemma (1.53), $\rho(X_s - X_s')$ is a scalar endomorphism of V_ρ. On the other hand, since $\mathfrak{g} = [\mathfrak{g}, \mathfrak{g}]$, $\text{trace}(\rho(X_s - X_s')) = 0$; thus $X_s - X_s' = 0$. This proves the second assertion in the corollary.

For $X \in \mathfrak{g}$, $\mathfrak{g}$ semi-simple, X_s and X_n, as in the corollary, are respectively the *semi-simple part* and *nilpotent part* of X.

7.11. Proposition. *Let $\mathfrak{g}$ be a semi-simple Lie algebra and $\mathfrak{h}$ a semi-simple subalgebra. Let $\mathfrak{c}^*(\mathfrak{h})$ be the centralizer of $\mathfrak{h}$ in $\mathfrak{g}$. Then $B_\mathfrak{g}|_{\mathfrak{c}^*(\mathfrak{h})}$ is non-degenerate (and hence $\mathfrak{c}^*(\mathfrak{h})$ is reductive (cf. 6.92)). Moreover, for every X in the centre of $\mathfrak{c}^*(\mathfrak{h})$, $ad(X)$ is semi-simple.*

7.12. Proof. The adjoint representation $ad_\mathfrak{g}$ restricted to $\mathfrak{h}$ is completely reducible (Corollary 7.7). Thus $\mathfrak{g}$ decomposes into a direct sum $\coprod_{i \in I} \mathfrak{g}_i$ of irreducible representations $\mathfrak{g}_i$ of $\mathfrak{h}$. Let $\sim$ be the equivalence relation on I defined as follows: for $i, j \in I$, $i \sim j$ if $\mathfrak{g}_i \simeq \mathfrak{g}_j$ or $\mathfrak{g}_j^*$ the dual of $\mathfrak{g}_j$. Let S be the set of equivalence classes of elements of I. For $s \in S$, let $\mathfrak{g}_s = \coprod_{i \in s} \mathfrak{g}_i$.

Now the non-degenerate form $B_\mathfrak{g}$ gives an isomorphism of $\mathfrak{g}$ on its dual as a $\mathfrak{g}$-module and hence also as a $\mathfrak{h}$-module. It follows from this that $B_\mathfrak{g}$ is non-degenerate on $\mathfrak{g}_s$ for all $s \in S$. Clearly, $\mathfrak{c}^*(\mathfrak{h}) = \mathfrak{g}_{s_o}$ where $s_o = \{i \in I | \mathfrak{g}_i \sim$ *the trivial $\mathfrak{h}$-module*$\}$. Thus $B_\mathfrak{g}|_{\mathfrak{c}^*(\mathfrak{h})}$ is non-degenerate.

Suppose now that X is an element of $\mathfrak{c}^*(\mathfrak{h})$. Let X_s and X_n be the semi-simple and nilpotent parts of X in $\mathfrak{g}$. Then since $ad(X_n) = ad(X)_n$, one sees that X_n is in the centre of $\mathfrak{c}^*(\mathfrak{h})$. It follows that $ad(X_n) \circ ad(Y)$ is nilpotent for all $Y \in \mathfrak{c}^*(\mathfrak{h})$ so that $B_\mathfrak{g}(X_n, Y) = 0$ for all $Y \in \mathfrak{c}^*(\mathfrak{h})$. Since $B_\mathfrak{g}$ is non-degenerate on $\mathfrak{c}^*(\mathfrak{h})$, this means that $X_n = 0$. Thus $X = X_s$ is semi-simple.

7.13. Theorem. *Let $\mathfrak{g}$ be a Lie algebra and $\mathfrak{r}$ its radical. Then $[\mathfrak{g}, \mathfrak{r}] = [\mathfrak{g}, \mathfrak{g}] \cap \mathfrak{r}$ and $\bar{\mathfrak{g}} = \mathfrak{g}/[\mathfrak{g}, \mathfrak{r}]$ is isomorphic to the direct product of its centre $\bar{\mathfrak{c}}(= \mathfrak{r}/[\mathfrak{g}, \mathfrak{r}])$ and the semi-simple algebra $\mathfrak{s} = [\bar{\mathfrak{g}}, \bar{\mathfrak{g}}](\simeq \mathfrak{g}/\mathfrak{r}$ under the natural map $\mathfrak{g} \to \mathfrak{g}/\mathfrak{r}$).*

7.14. Proof. Evidently $\bar{\mathfrak{r}} = \mathfrak{r}/[\mathfrak{g}, \mathfrak{r}]$ is the radical of $\bar{\mathfrak{g}}$. As $[\mathfrak{g}, \mathfrak{r}] \subset [\mathfrak{g}, \mathfrak{g}] \cap \mathfrak{r}$, we need to only show that $[\bar{\mathfrak{g}}, \bar{\mathfrak{g}}] \cap \bar{\mathfrak{r}} = \{0\}$. Now $\bar{\mathfrak{r}}$ is central in $\bar{\mathfrak{g}}$. Thus the adjoint representation of $\bar{\mathfrak{g}}$ factors through the semi-simple algebra $\mathfrak{g}/\mathfrak{r}$ and is therefore completely reducible. It follows that there is an ideal $\bar{\mathfrak{h}}$ in $\bar{\mathfrak{g}}$ such that $\bar{\mathfrak{g}} = \bar{\mathfrak{h}} \oplus \bar{\mathfrak{r}}$. As $\bar{\mathfrak{h}} \simeq \mathfrak{g}/\mathfrak{r}$, it is semi-simple and hence $[\bar{\mathfrak{g}}, \bar{\mathfrak{g}}] = [\bar{\mathfrak{h}}, \bar{\mathfrak{h}}] = \bar{\mathfrak{h}}$ and hence $[\bar{\mathfrak{g}}, \bar{\mathfrak{g}}] \cap \bar{\mathfrak{r}} = \{0\}$.

The next result indicates that the study of the structure of a Lie algebra essentially reduces to studying the structure of solvable and semi-simple algebras separately.

7.15. Theorem. *Let $\mathfrak{g}$ be a Lie algebra and $\mathfrak{r}$ its radical. Then there is a Lie subalgebra $\mathfrak{h}_o$ of $\mathfrak{g}$ such that $\mathfrak{h}_o \cap \mathfrak{r} = \{0\}$ and $\mathfrak{h}_o$ maps isomorphically onto $\mathfrak{h} = \mathfrak{g}/\mathfrak{r}$: in other words, $\mathfrak{g}$ is isomorphic to the semi-direct product of $\mathfrak{h}_o$ and $\mathfrak{r}$.*

A subalgebra $\mathfrak{h}_o$ of $\mathfrak{g}$ as above is called a *Levi subalgebra* of $\mathfrak{g}$.

7.16. Proof. We argue by induction on the length $ld(\mathfrak{r})$ of the derived series of $\mathfrak{r}$. Suppose that $ld(\mathfrak{r}) = 1$; this means that $\mathfrak{r}$ is abelian. Let σ denote the adjoint representation of $\mathfrak{g}$ on $\mathfrak{r}$ and set $\tau = \mathrm{Hom}_k(ad_\mathfrak{g}, \sigma)$: for $f : \mathfrak{g} \to \mathfrak{r}$, $X \in \mathfrak{g}$ and $v \in \mathfrak{g}$, $\tau(X)(f)(v) = \sigma(X)(f(v)) - f(ad(X)(v))$. Let $V = \{T \in V_\tau | \ T|_\mathfrak{r} \text{ is a scalar}\}$. Then V is stable under $\tau(\mathfrak{g})$. Evidently, $ad(\mathfrak{r}) \subset V$ (since $\mathfrak{r}$ is an ideal in $\mathfrak{g}$, $ad(X)(\mathfrak{g}) \subset \mathfrak{r}$ and $ad(X)(\mathfrak{r}) = \{0\}$ for all $X \in \mathfrak{r}$). It follows that $ad(\mathfrak{r})$ is a $\tau(\mathfrak{g})$-stable subspace of V. Let τ' denote the representation of $\mathfrak{g}$ on $V' = V/ad_\mathfrak{g}(\mathfrak{r})$.

The map $V \ni T \leadsto T|_\mathfrak{r}$ is easily seen to be a $\mathfrak{g}$-module morphism of V on to the trivial $\mathfrak{g}$-module. It is evidently 0 on $ad(\mathfrak{r})(\subset V)$ and hence defines a $\mathfrak{g}$-module morphism $\pi : V' \to k$. Now the representation τ', as is easily seen, factors through the semi-simple Lie algebra $\mathfrak{g}/\mathfrak{r}$ and is therefore completely reducible. Let $r : k \to V'$ be a $\mathfrak{g}$-module morphism such that $\pi \circ r = 1_k$ and let $T' = r(1)$. Then evidently

$$\tau'(X)(T') = 0, \ \forall \ X \in \mathfrak{g}. \tag{$*$}$$

Let $T \in V$ be such that its image in V' is T' Now $(*)$ implies that for every $X \in \mathfrak{g}$, there is a $Z(X) \in \mathfrak{r}$ such that

$$ad(X)_\circ T - T_\circ ad(X) = ad(Z(X)). \tag{1}$$

Since $\mathfrak{r}$ is abelian and the image of T lies in $\mathfrak{r}$, it follows that

$$ad(Z(X))_\circ T - T_\circ ad(Z(X)) = -ad(Z(X)). \tag{2}$$

(1) and (2) above imply that we have

$$ad(X + Z(X))_\circ T - T_\circ ad(X + Z(X)) = 0. \tag{3}$$

Thus $X = (X + Z(X)) - Z(X)$ leading to the conclusion that

$$\mathfrak{g} = \mathfrak{r} + \mathfrak{h}$$

where $\mathfrak{h} = \{X \in \mathfrak{g}| \ \tau(X)(T) = 0\}$. Consider now the exact sequence

$$\{0\} \to \mathfrak{h} \cap \mathfrak{r} \to \mathfrak{h} \to \mathfrak{g}/\mathfrak{r} \to \{0\}.$$

If $Z \in \mathfrak{h} \cap \mathfrak{r}$, we have

$$0 = ad(Z)_\circ T - T_\circ ad(Z) = -T_\circ ad(Z) = -ad(Z).$$

Thus $\mathfrak{h} \cap \mathfrak{r}$ is in the centre of $\mathfrak{g}$. The adjoint action of $\mathfrak{h}$ on itself factors through the semi-simple algebra $\mathfrak{h}/\mathfrak{h} \cap \mathfrak{r}$ ($\simeq \mathfrak{g}/\mathfrak{r}$) and is hence completely reducible. Hence there exists an ideal $\mathfrak{h}'$ in $\mathfrak{h}$ such that $\mathfrak{h} = \mathfrak{h}' \oplus (\mathfrak{r} \cap \mathfrak{h})$. This means that the subalgebra $\mathfrak{h}' = [\mathfrak{h}, \mathfrak{h}]$ maps isomorphically on to $\mathfrak{h}/\mathfrak{h} \cap \mathfrak{r}$. This takes care of the start of the induction.

Assume that the theorem is proved when $ld(\mathfrak{r}) \leq m, m \geq 1$. Let $ld(\mathfrak{r}) = m + 1$. Then $\mathfrak{a} = D^m(\mathfrak{r})$ is an abelian ideal in $\mathfrak{g}$. Set $\mathfrak{g}' = \mathfrak{g}/\mathfrak{a}$ and $\mathfrak{r}' = \mathfrak{r}/\mathfrak{a}$; then $\mathfrak{r}'$ is the radical of $\mathfrak{g}'$ and has derived series length m. By the induction hypothesis, we can find a subalgebra $\mathfrak{h}'_o$ of $\mathfrak{g}'$ which maps isomorphically onto $\mathfrak{g}'/\mathfrak{r}'$ ($= \mathfrak{g}/\mathfrak{r} = \mathfrak{h}$). Let $\pi : \mathfrak{g} \to \mathfrak{g}'$ be the natural map and set $\mathfrak{g}'' = \pi^{-1}(\mathfrak{h}'_o)$; then the radical of $\mathfrak{g}''$ is $\mathfrak{a}$. By induction hypothesis again, we can find a subalgebra $\mathfrak{h}_o$ of $\mathfrak{g}''$ which maps isomorphically onto $\mathfrak{g}''/\mathfrak{a}$. Clearly $\mathfrak{h}_o$ is a subalgebra which maps isomorphically on to $\mathfrak{h}$ under the map $\mathfrak{g} \to \mathfrak{h}$.

7.17. We will now prove a uniqueness assertion about the subalgebra $\mathfrak{h}_o$ in the above theorem. Set $[\mathfrak{g}, \mathfrak{r}] = \mathfrak{n}_o$. Then $\mathfrak{n}_o$ is contained in $\mathfrak{n}$, the maximum nilpotent ideal in $\mathfrak{g}$. Now if $X \in \mathfrak{n}$, $ad(X)$ is nilpotent and hence $\exp(ad(X)) = \sum_{r=0}^{\infty} ad(X)^r/r! = \sum_{r=0}^{d_\mathfrak{g}} ad(X)^r/r!$ (recall that $d_\mathfrak{g} = \dim_k(\mathfrak{g})$) is an automorphism of the Lie algebra $\mathfrak{g}$. Now if $X, Y \in \mathfrak{n}$ are such that $[X, Y] = 0$, then $\exp(ad(X + Y)) = \exp(ad(X)) \cdot \exp(ad(Y))$; in particular $\exp(ad(X)) \cdot \exp(ad(-X)) = 1_\mathfrak{g}$ and hence $\exp(ad(X))$ is an automorphism of the vector space $\mathfrak{g}$. In fact it is a Lie algebra automorphism as can be seen easily (using the fact that $ad(X)$ is a derivation). With this background we will now prove

7.18. Theorem (Malcev). *Let $\mathfrak{g}$ be a Lie algebra, $\mathfrak{r} = \mathfrak{r}(\mathfrak{g})$ its radical and $\pi : \mathfrak{g} \to \mathfrak{g}/\mathfrak{r}(= \mathfrak{h})$ be the natural map. Suppose $\mathfrak{h}_i, i = 1, 2$ are subalgebras such that $\pi|_{\mathfrak{h}_i}$ is an isomorphism of $\mathfrak{h}_i$ on $\mathfrak{h}$. Then there exists $X \in \mathfrak{n}_o (= \mathfrak{n}_o(\mathfrak{g}) = [\mathfrak{g}, \mathfrak{r}])$ such that $\exp(ad(X)) \circ r_1 = r_2$ where $r_i : \mathfrak{h} \to \mathfrak{g}$ is the inclusion of $\mathfrak{h}_i$ in $\mathfrak{g}$ composed with the Lie algebra isomorphism $(\pi|_{\mathfrak{h}_i})^{-1} : \mathfrak{h} \to \mathfrak{h}_i$.*

7.19. Proof. If $\mathfrak{n}_o = 0$, $\mathfrak{r}$ is central in $\mathfrak{g}$ and $\mathfrak{g} = [\mathfrak{g}, \mathfrak{g}] \times \mathfrak{c}(\mathfrak{g})$. It follows that $[\mathfrak{g}, \mathfrak{g}]$ is the unique subalgebra of $\mathfrak{g}$ that maps isomorphically onto $\mathfrak{h}$. We assume then that $\mathfrak{n}_o \neq \{0\}$. We will prove the theorem by induction on the length $\lambda(\mathfrak{n}_o)$ of the ascending central series of $\mathfrak{n}_o$. Let $\mathfrak{c}(\mathfrak{n}_o)$ be the centre of $\mathfrak{n}_o$. Since $\mathfrak{n}_o \neq \{0\}$, $\mathfrak{c}(\mathfrak{n}_o) \neq 0$. It is an ideal in $\mathfrak{g}$. Set $\mathfrak{g}' = \mathfrak{g}/\mathfrak{c}(\mathfrak{n}_o)$ and let $p : \mathfrak{g} \to \mathfrak{g}'$ be the natural map. Then $p(\mathfrak{r}) = \mathfrak{r}'(= \mathfrak{r}(\mathfrak{g}'))$ is the radical of $\mathfrak{g}'$; evidently $\mathfrak{n}_o' \overset{\text{def}}{=} \pi(\mathfrak{n}_o) = [\mathfrak{g}', \mathfrak{r}'] = \mathfrak{n}_o(\mathfrak{g}')$ and the length $\lambda(\mathfrak{n}_o')$ of the ascending central series of $\mathfrak{n}_o'$ is $\lambda(\mathfrak{n}_o) - 1$. So if the result is proved for all Lie algebras $\mathfrak{b}$ with $[\mathfrak{b}, \mathfrak{r}(\mathfrak{b})]$ of ascending central series length less than $\lambda(\mathfrak{n}_o)$, we can find $X' \in \mathfrak{n}_o'$ such that $\exp(ad(X')) \circ p \circ r_1 = p \circ r_2$. Let $\tilde{X}' \in \mathfrak{n}_o$ be such that $p(\tilde{X}') = X'$. Let $\mathfrak{h}_3 = \{\exp(ad(-\tilde{X}'))\}(\mathfrak{h}_2)$. Then $p(\mathfrak{h}_3) = p(\mathfrak{h}_1)$ so that $\mathfrak{h}_3$ and $\mathfrak{h}_1$ are subalgebras of $p^{-1}(\mathfrak{h}_1)$ which map isomorphically onto $p^{-1}(\mathfrak{h}_1)/\mathfrak{c}(\mathfrak{n}_o)$. It follows by the induction hypothesis that there is a $Y \in \mathfrak{c}(\mathfrak{n}_o)$ such that $\exp(ad(-\tilde{X}')) \circ r_2 = r_3 = \{\exp(ad(Y)) \circ r_1$. Thus $r_2 = \exp(ad(\tilde{X}')) \circ \exp(ad(Y)) \circ r_1 = \exp(ad(\tilde{X}' + Y)) \circ r_1$ (since $[\tilde{X}', Y] = 0$). Hence the theorem.

We now classify Lie algebras of dimension ≤ 3.

7.20. Theorem. *A Lie algebra of dimension 1 is abelian. A Lie algebra of dimension 2 is either abelian or isomorphic to the semi-direct product $k \propto k$ where elements of the first factor act on the second by multiplication. A Lie algebra of dimension 3 is isomorphic to one of the following five Lie algebras:*
(i) The abelian Lie algebra k^3.
(ii) The direct product of k and $k \propto k$ as above.
(iii) The Lie subalgebra of upper triangular nilpotent matrices in $M(3, k)$. This Lie algebra is called the Heisenberg Lie algebra.
(iv) Let $\sigma : k \to \mathfrak{gl}(2, k)$ be a representation of k and $\mathfrak{g}_\sigma = k \propto_\sigma k^2$. When σ is trivial $\mathfrak{g}$ is abelian and one recovers (i). When σ is a direct sum of two 1-dimensional representations one of which is trivial and the other τ non-trivial, then $\mathfrak{g} \simeq k \times (k \propto_\tau k)$. If $\sigma(1)$ is nilpotent and $\neq 0$, $\mathfrak{g}_\sigma$ is isomorphic to the Heisenberg Lie algebra. In general, $\mathfrak{g}_\sigma \simeq \mathfrak{g}_{\sigma'}$ if and only if σ is equivalent to σ'.
(v) The Killing form $B_\mathfrak{g}$ on the 3-dimensional algebra $\mathfrak{g}$ is non-degenerate and the adjoint representation gives an isomorphism of $\mathfrak{g}$ on the orthogonal Lie algebra $\mathfrak{o}(B) = \{X \in \mathfrak{g}| \ B_\mathfrak{g}([X, A], B) + B_\mathfrak{g}(A, [X, B]) = 0, \ \forall A, B \in \mathfrak{g}\}$.

7.21. Proof. The first two assertions are left as exercises to the reader. Let $\mathfrak{g}$ be a Lie algebra of dimension 3. If the Killing form is non-degenerate, one checks easily that the adjoint representation yields an isomorphism of $\mathfrak{g}$ on $\mathfrak{o}(B_{\mathfrak{g}})$. When the Killing form is identically 0, $\mathfrak{g}$ is solvable and it is not difficult to see that $\mathfrak{g}$ is either abelian or isomorphic to the Heisenberg Lie algebra. When the kernel $\mathfrak{a}$ of the Killing form is of dimension 1, the Killing form of $\mathfrak{g}/\mathfrak{a}$ is non-degenerate which contradicts the classification of 2-dimensional Lie algebras. Hence the kernel of the Killing form, when it is degenerate, is of dimension 2 and is hence necessarily abelian. Pick an element $X \in \mathfrak{g} - \mathfrak{a}$. Then we see that $\mathfrak{g}$ is a semi direct product $((k \simeq) k \cdot X) \propto_{\sigma} \mathfrak{a}$ where σ is the adjoint representation of $k \cdot X$ on $\mathfrak{a}$. This completes the proof of the theorem.

7.22. The Lie Algebra $\mathfrak{sl}(2,k)$. Set

$$e = \begin{pmatrix} 0 & 1 \\ 0 & 0 \end{pmatrix} \quad h = \begin{pmatrix} 1 & 0 \\ 0 & -1 \end{pmatrix} \quad f = \begin{pmatrix} 0 & 0 \\ 1 & 0 \end{pmatrix}.$$

Then $\{e, h, f\}$ is a basis of $\mathfrak{sl}(2, k)$ (which in the sequel, will be referred to as the *standard basis* of $\mathfrak{sl}(2, k)$). We have then

$[h, e] = 2 \cdot e$, $[h, f] = -2 \cdot f$ and $[e, f] = h$.

The eigen-values of $ad_{\mathfrak{sl}(2,k)}(h)$ are obviously $\{-2, 0, 2\}$, $\{f, h, e\}$ being the respective eigen-vectors for them. Now if $\mathfrak{a} \neq \{0\}$ is an ideal in $\mathfrak{sl}(2, k)$, it is stable under $ad_{\mathfrak{sl}(2,k)}(h)$ and therefore contains one of $\{e, h, f\}$; and it is a simple exercise to check that any one of $\{e, h, f\}$ generates $\mathfrak{sl}(2, k)$ as an ideal. Thus $\mathfrak{sl}(2, k)$ has no proper ideals and is therefore a simple Lie algebra. Also one has

$$B_{\mathfrak{sl}(2,k)}(h, h) = 2 \cdot B_{\mathfrak{sl}(2,k)}(e, f) = 8 \text{ while}$$

$$B_{\mathfrak{sl}(2,k)}(e, e) = B_{\mathfrak{sl}(2,k)}(f, f) = B_{\mathfrak{sl}(2,k)}(h, f) = B_{\mathfrak{sl}(2,k)}(h, e) = 0.$$

Hence the Killing form $B_{\mathfrak{sl}(2,k)}$ of $\mathfrak{sl}(2, k)$ is *non-degenerate* and $\mathfrak{sl}(2, k) \simeq \mathfrak{o}(B_{\mathfrak{sl}(2,k)})$. Recall that a symmetric bilinear B form on a vector space V is *isotropic* if there is a $v \neq 0$ in V such that $B(v, v) = 0$. Evidently $B_{\mathfrak{sl}(2,k)}$ is isotropic. It is well known – and easy to see – that any two non-degenerate isotropic bilinear forms in three variables are equivalent. Thus $\mathfrak{sl}(2, k)$ is isomorphic to the orthogonal Lie algebra $\mathfrak{o}(B)$ of a non-degenerate isotropic symmetric bilinear form B on k^3.

7.23. Representations of $\mathfrak{sl}(2, k)$. Let ρ be an irreducible representation of $\mathfrak{sl}(2, k)$ and let $\bar{\rho}$ be its extension to $\mathfrak{sl}(2, \bar{k})(= \mathfrak{sl}(2, k) \otimes_k \bar{k})$ (so that $V_{\bar{\rho}} = V_{\rho} \otimes_k \bar{k} = \bar{V}_{\rho}$). Set $V = V_{\rho}$ and $\bar{V} = \bar{V}_{\bar{\rho}}$. Let $\mathfrak{E}(\rho(h))$ be the set of eigenvalues of $\rho(h)$ and, for $\mu \in \mathfrak{E}(\rho(h))$, let $\bar{V}(\mu, h)$ be the eigen-space of $\rho(h)$ corresponding to μ. Then

$$\rho(e)(\bar{V}(\mu, h)) \subset \bar{V}(\mu + 2, h) \text{ and } \rho(f)(\bar{V}(\mu, h)) \subset \bar{V}(\mu - 2, h). \quad (*)$$

This means that $\rho(e)$ and $\rho(f)$ are nilpotent. Set $V^e = \{v \in V \mid \rho(e)(v) = 0\}$, $\bar{V}^e = \{v \in \bar{V} \mid \rho(e)(v) = 0\}$ and $V^f = \{v \in V \mid \rho(f)(v) = 0\}$, $\bar{V}^f = \{v \in \bar{V} \mid \rho(f)(v) = 0\}$. Then $\bar{V}^e$ (resp. $\bar{V}^f$) is the $\bar{k}$-linear span of V^e (resp. V^f) in $\bar{V}$.

Evidently, $\bar{W} = \coprod_{n \in \mathbb{Z}} \bar{V}(\mu - 2 \cdot n, h)$ is $\rho(h)$ stable. From $(*)$ it is clear that $\bar{W}$ is stable under $\rho(e)$ and $\rho(f)$ as well, hence under $\mathfrak{sl}(2, \bar{k})$. Let $\bar{\sigma}$ denote the sub-representation of $\mathfrak{sl}(2, \bar{k})$ on $\bar{W}$. Then we have

$$0 = \operatorname{trace}([\bar{\sigma}(e), \bar{\sigma}(f)]) = \operatorname{trace}(\sigma(h)) = \sum_{n \in \mathbb{N}} \dim_k \bar{V}_{\bar{\rho}}(\mu - 2 \cdot n, h)(\mu - 2 \cdot n)$$

leading to the conclusion that $\mu \in \mathbb{Q}(\subset k)$. Since $\mu \in \mathfrak{E}(\rho(h))$ is arbitrary, $\mathfrak{E}(\rho(h)) \subset k$. Thus we see that for every $\mu \in \mathfrak{E}(\rho(h))$, $\bar{V}(\mu, h) = V(\mu, h) \otimes_k \bar{k}$ where $V(\mu, h) = \{v \in V_\rho \mid \rho(h)(v) = \mu \cdot v\}$.

Let $v_0 \in V^e, v_0 \neq 0$ be an eigen-vector for $\rho(h)$ for an eigen-value λ. For $q \in \mathbb{N}$, set $v_q = \rho(f)^q(v_0)$ and let d be the least integer such that $\rho(f)^d(v_q) = 0$. Then as the $\{v_q \mid 0 \leq q < d\}$ are non-zero and belong to distinct eigen-spaces for $\rho(h)$, they are linearly independent. A simple induction on q using the fact that $[e, f] = h$ leads to

$$\rho(e)(v_q) = q(\lambda - (q-1)) \cdot v_{(q-1)} \qquad (**)$$

for $0 \leq q \leq d$. It follows from $(**)$ that the k-linear span W of $\{v_q \mid 0 \leq q \leq (d-1)\}$ is $\mathfrak{sl}(2, k)$-stable; and since ρ is irreducible, $W = V$. Since $\dim_k W = d$, we conclude that $d_\rho = \dim_k V = d$. Taking $q = d$ in $(**)$, we see $\lambda = (d-1)$. We have thus proved the following.

7.24. Theorem. *For each integer $n \geq 1$, $\mathfrak{sl}(2, k)$ admits an n-dimensional irreducible representation $\rho(n)$ satisfying the following conditions:*
$V_{\rho(n)}$ admits a basis $\{v_q \mid 0 \leq q \leq (n-1)$ (so that $d_{\rho(n)} = n$) such that
(i) $\rho(n)(f)(v_q) = v_{(q+1)}$ for $0 \leq q \leq (d_\rho - 1)$ and $\rho(n)(f(v_{(d_\rho-1)})) = 0$.
(ii) $\rho(n)(h)(v_q) = (n - 1 - 2 \cdot q) \cdot v_q$.
(iii) $\rho(n)(e)(v_q) = q \cdot (n - q) \cdot v_{(q-1)}.$ —
Any irreducible representation ρ with $d_\rho = n$ is isomorphic to $\rho(n)$. Also, $\rho(n) \simeq S^{n-1}(\nu)$, the symmetric $(n-1)^{th}$ power of ν where ν is the natural representation $\rho(2)$ of $\mathfrak{sl}(2, k)$ (see 1.47 (2)).

The last assertion is left as an exercise to the reader.

7.25. Corollary. *Let ρ, ρ' be representations of $\mathfrak{sl}(2, k)$. Then the following are equivalent:*
(i) $\rho \simeq \rho'$, (ii) $\rho|_{k \cdot e} \simeq \rho'|_{k \cdot e}$, (iii) $\rho|_{k \cdot h} \simeq \rho'|_{k \cdot h}$, (iv) $\rho|_{k \cdot f} \simeq \rho'|_{k \cdot f}$.

This is an immediate consequence of Theorem 7.24 in the light of complete reducibility of representations of $\mathfrak{sl}(2, k)$. Note that there is an automorphism of $\mathfrak{sl}(2, k)$ taking e (resp. f) into f (resp. e) and h to $-h$, viz. conjugation by the matrix $w = \begin{pmatrix} 0 & 1 \\ 1 & 0 \end{pmatrix}$.

7.26. Corollary. *Given any nilpotent matrix X in $M(n, k)$, there is a representation ρ (resp. ρ' of $\mathfrak{sl}(2, k) \to M(n, k)$ such that $\rho(f) = X$ (resp. $\rho'(e) = X$).*

7.27. Proof. This is just a reformulation of the reduction of a nilpotent matrix to the Jordan canonical form: given any partition $P : n = \sum_{i=1}^{r} n_i$, let $\rho_P = \coprod_{i=1}^{r} \rho(n_i)$; then $\rho_P(f)$ referred to the basis of k^n obtained from the bases of the V_{n_i} in Theorem 7.24 is evidently in Jordan canonical form; and every nilpotent matrix X in Jordan canonical form determines a partition of $n = \sum_{i=1}^{r} n_i$ such that $X = \rho(f)$ where $\rho = \coprod_{i=1}^{r} \rho(n_i)$.

The above corollary has the following generalization which is useful in the study of finer structural properties of Lie algebras.

7.28. Theorem (Jacobson, Morozov). *Let $\mathfrak{g}$ be semi-simple and $X \in \mathfrak{g}$ be such that $ad(X)$ is nilpotent. Then there is a homomorphism $u : \mathfrak{sl}(2, k) \to \mathfrak{g}$ with $u(e) = X$. Let $\mathfrak{z}(X)$ be the centralizer of X in $\mathfrak{g}$. Suppose that $H' \in ad(X)(\mathfrak{g})$ is such that $[H', X] = 2 \cdot X$. Then there is an element $A \in \mathfrak{n}(\mathfrak{z}(X))$ (the nil-radical of $\mathfrak{c}(X)$) such that $ad_{\mathfrak{g}}(X)$ is nilpotent and $\exp(ad(A)(H')) = H$.*

7.29. Proof. In the special case $\mathfrak{g} = \mathfrak{sl}(n, k)$, the first assertion of the theorem is nothing but Corollary 7.26. We will now show that if the first assertion holds for a semi-simple $\mathfrak{g}$, the second assertion holds for it as well. Set $\mathfrak{h} = \mathfrak{sl}(2, k)$. Fix a homomorphism $u : \mathfrak{h} \to \mathfrak{g}$ with $u(e) = X$. Then we have the following description of the centralizer $\mathfrak{z}(X)$ of X in $\mathfrak{g}$: under $\rho = ad_{\mathfrak{g}} \circ u$, $\mathfrak{g}$ decomposes into a direct sum $\coprod_{q \in \mathbb{N}} W(q)$ where for $q \in \mathbb{N}$, $W(q)$ is the sum of all the irreducible $\mathfrak{h}$-sub-representations of $\mathfrak{g}$ of dimension q. Let $Z(q) = \{w \in W(q)| \ ad_{\mathfrak{g}}(u(h))(w) = (q-1) \cdot w\}$. Then $Z(q) \subset \mathfrak{z}(X)$ and

$$\mathfrak{z}(X) = \sum_{q \in \mathbb{N}} Z(q)$$

(see 7.35). Now if $E(r)$ is the eigen-space of $ad_{\mathfrak{g}} u(h)$ corresponding to an eigen-value r, one has for $r, s \in \mathbb{N}$, $[E(r), E(s)] \subset E(r + s)$. It follows from this that $\mathfrak{n} = \sum_{q \in \mathbb{N}, q \geq 2} Z(q)$ is an ideal in $\mathfrak{z}(X)$ and that $ad_{\mathfrak{g}}(A)$ is nilpotent for every $A \in \mathfrak{n}$. It is thus contained in the maximum nilpotent ideal of $\mathfrak{z}(X)$. Moreover $H = \rho(h)$ normalizes $\mathfrak{n}$ and for $r \geq 2$, $\mathfrak{n}_r = \sum_{q \geq r} Z(q)$, is an ideal in $\mathfrak{n}$. Let $\mathfrak{b} = k \cdot H \oplus \mathfrak{n}$. As H stabilizes the $Z(n)$ for every n, $\mathfrak{b}$ is a solvable subalgebra of $\mathfrak{g}$ and $\mathfrak{n}_q$ is an ideal in $\mathfrak{b}$ for all $q \geq 2$. Also $\mathfrak{n}_q = \{0\}$ for all large q. Since $H - H'$ is in the image of $ad_{\mathfrak{g}}(X)$, from the representation theory of $\mathfrak{sl}(2,k)$ (7.35), we see that $H'-H \in \mathfrak{n}$ so that $H' \in \mathfrak{b}$. For $q \geq 2$ let A_q be the following assertion:

A_q: *Let H' be an element in $\mathfrak{b}$ such that $H-H' \in \mathfrak{n}_q$. Then there is an element $Y_q \in \mathfrak{n}$ such that $\exp_{\mathfrak{b}}(ad_{\mathfrak{g}}(Y_q))(H') = H$.*

Evidently, A_2 yields the second statement of the theorem. We will prove A_q by downward induction on q. For large q, $\mathfrak{n}_q = \{0\}$ and so A_q holds trivially. Assume that A_q holds for all $q \geq p(> 2)$. We will then prove that A_{p-1} holds. Let $\bar{\mathfrak{b}} = \mathfrak{b}/\mathfrak{n}_p$, $\bar{\mathfrak{n}} = \mathfrak{n}/\mathfrak{n}_p$, $\bar{\mathfrak{n}}_{p-1} = \mathfrak{n}_{p-1}/\mathfrak{n}_p$ and $\bar{H}$ (resp. $\bar{H}'$) the image of H (resp. H') in $\bar{\mathfrak{b}}$ under the natural map $\mathfrak{b} \to \bar{\mathfrak{b}}$. Then $\bar{\mathfrak{n}}_{p-1}$ is an abelian ideal in $\bar{\mathfrak{b}}$ and (as is easy to see) for any $X \in \bar{\mathfrak{n}}_{p-1}$, $ad_{\bar{\mathfrak{b}}}(X)^2 = 0$. It follows that if $H-H' \in \mathfrak{n}_{p-1}$, $\exp(ad_{\bar{\mathfrak{b}}}(\bar{H}-\bar{H}'))(\bar{H}') = \bar{H}'+(\bar{H}-\bar{H}') = \bar{H}$. Now let $H'' = \exp(ad_{\mathfrak{b}}(H-H'))(H')$. Then $H-H'' \in \mathfrak{n}_p$; and since A_p holds, there is a $Y \in \mathfrak{n}_p$ such that $\exp(ad_{\mathfrak{b}}(Y))(H'') = H$. Thus $\exp(ad_{\mathfrak{b}}(Y))(\exp(ad_{\mathfrak{b}}(H-H'))(H')) = H$. Now since $ad_{\mathfrak{b}}(\mathfrak{n})$ consists entirely of nilpotent endomorphisms we have, setting $g = \exp(ad_{\mathfrak{b}}(Y)) \circ \exp(ad_{\mathfrak{b}}(H-H'))$ and $Y_p' = \ln(1 + (g-1)) = \sum_{q \in \mathbb{N}} (-1)^q \cdot (g-1)^{q+1}$ (note that this is a finite sum), $g = \exp(Y_p')$ and $Y_{p-1}' = ad_{\mathfrak{b}}(Y_{p-1})$ for a suitable $Y_{p-1} \in \mathfrak{n}$. Hence A_{p-1} holds. Thus we see the second assertion of the theorem holds, if the first does.

We will now show that the first assertion holds for any semi-simple $\mathfrak{g}$. We have seen that it holds for $\mathfrak{g} = \mathfrak{gl}(\mathfrak{g})$. Let $\mathfrak{g}$ be any semi-simple Lie algebra. The adjoint representation then gives a realization of $\mathfrak{g}$ as a Lie subalgebra of $\mathfrak{gl}(\mathfrak{g})$. The adjoint representation of $\mathfrak{gl}(\mathfrak{g})$ restricted to $\mathfrak{g}$ being completely reducible, we have a $\mathfrak{g}$-submodule E of $\mathfrak{gl}(\mathfrak{g})$ such that $\mathfrak{gl}(\mathfrak{g}) = \mathfrak{g} \oplus E$. Suppose now that $X \in \mathfrak{g}$ is a nilpotent element. Then there is a morphism $u : \mathfrak{sl}(2,k) \to \mathfrak{gl}(\mathfrak{g})$ such that $u(e) = X$. Set $H = u(h)$ and $Y = u(f)$. Let $H = H_o + H'$ and $Y = Y_o + Y'$ with $H_o, Y_o \in \mathfrak{g}$ and $H', Y' \in E$. Then since $\mathfrak{g}$ and E are $ad_{\mathfrak{gl}(\mathfrak{g})}$-stable, we see that $[H_o, X] = 2 \cdot X$. Moreover from the fact that $[X,Y] = H$, we see that $[X, Y_o] = H_o$. Evidently $H - H_o \in ad_{\mathfrak{gl}(\mathfrak{g})} \cap \mathfrak{z}$, where $\mathfrak{z}$ is the centralizer of X in $\mathfrak{gl}(\mathfrak{g})$. It follows that there is an invertible element $T \in \mathfrak{z}$ such that $THT^{-1} = H_o$. Let $TYT^{-1} = Y^* = Y_o^* + Y^{*'}$ with $Y_o^* \in \mathfrak{g}$ and $Y^{*'} \in E$. Then one has from $[H_o, Y^*] = -2 \cdot Y^*$ and $[X, Y^*] = H_o$, $[H_o, Y_o^*] = -2 \cdot Y_o^*$ and $[X, Y^*] = H_o$. Clearly then the linear map $u' : \mathfrak{sl}(2,k) \to \mathfrak{g}$ defined by setting $u'(e) = X$, $u'(h) = H_o$ and $u'(f) = Y_o^*$ is a Lie algebra morphism with $u'(e) = X$ completing the proof of the theorem.

We end this chapter and the book with the following famous result.

7.30. Theorem (I D Ado). *Let $\mathfrak{g}$ be a (finite dimensional) Lie algebra over k. Then there is a faithful representation $\rho : \mathfrak{g} \to \mathfrak{gl}(V)$ of $\mathfrak{g}$ on some finite dimensional vector space.*

7.31. The Enveloping Algebra of $\mathfrak{g}$. For the proof of Ado's theorem we need some results relating to the enveloping algebra $U(\mathfrak{g})$ of $\mathfrak{g}$ (see 1.43) which is the quotient of the tensor algebra $T(\mathfrak{g})$ of $\mathfrak{g}$ by the 2-sided ideal $I_\mathfrak{g}$ generated by $\{X \otimes Y - Y \otimes X - [X,Y]| \ X, Y \in \mathfrak{g}\}$. One has an ascending filtration $\{U_p(\mathfrak{g})| \ p \in \mathbb{N}\}$ of $U(\mathfrak{g})$ by linear subspaces defined as follows: $U_p(\mathfrak{g})$ is the image of $\sum_{\{i \in \mathbb{N}| \ i \leq p\}} \otimes^i \mathfrak{g}\}$ under the natural map $\pi : T(\mathfrak{g}) \to U(\mathfrak{g})$. It is then obvious that for $a \in U_p(\mathfrak{g})$ and $b \in U_q(\mathfrak{g})$, we have $a \cdot b \in U_{(p+q)}(\mathfrak{g})$. It follows that we have a natural structure of a graded algebra on $\coprod_{p=0}^{\infty} U_p(\mathfrak{g})/U_{(p-1)}(\mathfrak{g})$ where we have set $U_{-1}(\mathfrak{g}) = \{0\}$: for $x \in U_p(\mathfrak{g})$ and $y \in U_q(\mathfrak{g})$, $\pi_p(x) \cdot \pi_q(y) = \pi_{(p+q)}(x \cdot y)$ where for $r \in N$, π_r is the natural map of $U_r(\mathfrak{g})$ on $U_r(\mathfrak{g})/U_{(r-1)}(\mathfrak{g}) \overset{\text{def}}{=} E_0^r(\mathfrak{g})$. A central result about the enveloping algebra is the following.

7.32. Theorem. *The natural map $T(\mathfrak{g}) \to U(\mathfrak{g})$ restricted to $\mathfrak{g}$ $(= \otimes^1 \mathfrak{g})$ maps $\mathfrak{g}$ injectively into $U_1(\mathfrak{g})$. The composite of the natural map π_1 with this injection is an isomorphism u of $\mathfrak{g}$ on $E_0^1(\mathfrak{g})$. The algebra $\sum_{i \in \mathbb{N}} E_0^i(\mathfrak{g}) \overset{\text{def}}{=} E_0(U(\mathfrak{g}))$ is commutative and the inclusion u induces an isomorphism of the symmetric algebra $S(\mathfrak{g})$ of $\mathfrak{g}$ on $E_0(U(\mathfrak{g}))$.*

7.33. Proof. Let $B = \{X_i| \ 1 \leq i \leq m\}$ be an ordered basis of $\mathfrak{g}$ (over k). For an integer $p \in \mathbb{N}$, let $\mathbf{B}(p)$ be the set of all maps $f : [1,p] \to B$ and for $f \in \mathbf{B}(p)$, $p \geq 1$ let

$$X_f = f(1) \otimes f(2) \otimes \cdots f(p-1) \otimes f(p) \in T^P(\mathfrak{g}).$$

When $p = 0$, $[1,p] = \phi$ and $\mathbf{B}(0)$ is the singleton set consisting of the unique inclusion $j\phi \subset B$ and we set $X_j = 1 (\in T^0(\mathfrak{g}))$. Then it is not difficult to see that the set

$$\{X_f \otimes \{(X_i \otimes X_j) - (X_j \otimes X_i) - u([X_i, X_j])\} \otimes X_g|$$
$$f \in \mathbf{B}(p), g \in \mathbf{B}(q), p, q \in \mathbb{N}, 1 \leq i < j \leq m\}$$

is a basis over k of the two-sided ideal $I_\mathfrak{g}$. It follows easily from this that $(\mathfrak{g} = T^1(\mathfrak{g})) \cap I_\mathfrak{g} = \{0\}$. This proves the first assertion of the theorem.

7.34. Proof of Theorem 7.32 (Continued). Next we note that $E_0^1(\mathfrak{g})$ generates $E_0(U(\mathfrak{g}))$ as an algebra. Since $\pi_1(X) \cdot \pi_1(Y) - \pi_1(Y) \cdot \pi_1(X) = 0$ for all $X, Y \in \mathfrak{g}$, $E_0(U(\mathfrak{g}))$ is commutative. By the universal property of the symmetric algebra we see that u extends to a surjective homomorphism of $S(\mathfrak{g})$ on $E_0(U(\mathfrak{g}))$. We observe that the diagonal map Δ maps $U_p(\mathfrak{g})$ (cf. 1.43) into $U_p(\mathfrak{g} \times \mathfrak{g}) = \sum_{r,s \in \mathbb{N}. r+s=p} U_r(\mathfrak{g}) \otimes_k U_s(\mathfrak{g})$ and hence defines an associative algebra morphism $E_0(\Delta) : E_(U(\mathfrak{g})) \to E_0(U(\mathfrak{g})) \otimes_k E_0(\mathfrak{g})$. One has evidently $E_0(\Delta)(X) = X \otimes 1 + 1 \otimes X$ for all $X \in E_0^1(U(\mathfrak{g}))$. Identify $E_0^1(U(\mathfrak{g}))$ with $\mathfrak{g}$ so that B may be regarded as a basis of $E_0^1(U(\mathfrak{g}))$. For $\alpha \in \mathbb{N}^m$ let $X^\alpha = X_1^{\alpha_1} \cdot X_2^{\alpha_2} \cdots X_m^{\alpha_m} (\in E_0^{|\alpha|}(U(\mathfrak{g})))$. We will show that $\{X^\alpha | \ \alpha \in \mathbb{N}^m, |\alpha| = p\}$ is a basis of $E_0^p U(\mathfrak{g})$. This is done by induction on p. The start of the induction at $p = 1$ is equivalent to the first assertion. Assume then that we have proved the assertion for $p < r$. The set $\{X^\alpha | \ |\alpha| = r\}$ evidently spans $E_0^r(U(\mathfrak{g}))$. Suppose now that we have a linear relation $\sum_{\alpha \in \mathbb{N}^m | \ |\alpha| = r} a_\alpha \cdot X^\alpha = 0$. Then one has $0 = \Delta(\sum_{\alpha \in \mathbb{N}^m | \ |\alpha| = r} a_\alpha \cdot X^\alpha)$. Now, since $\Delta(X^\alpha) = \prod_{1 \le i \le m} (\sum_{0 \le j \le \alpha_i} \binom{\alpha_i}{j} X_i^j \otimes X_i^{(\alpha_i - j)})$, the above linear relation leads to: $\sum_{\alpha \in \mathbb{N}^m, |\alpha| = r} (\sum_{\beta, \gamma \in \mathbb{N}^n, \beta + \gamma = \alpha, |\beta|, |\gamma| < r} a_\alpha \cdot \nu(\beta, \gamma) \cdot X_\beta \otimes X_\gamma) = 0$ where $\nu(\beta, \gamma)$ are positive integers. But this contradicts the linear independence of the set $\{X_\beta \otimes X_\gamma | \ \beta, \gamma \in \mathbb{N}^m, |\beta|, |\gamma| < r\}$ which is guaranteed by the induction hypothesis. This implies that u induces an isomorphism of $S(\mathfrak{g})$ on $E_0(U(\mathfrak{g}))$. Hence the theorem.

7.35. Suppose now that $\mathfrak{g}$ is the semi-direct product of a Lie subalgebra $\mathfrak{a}$ and an ideal $\mathfrak{r}$ in $\mathfrak{g}$. Then we have an action $\sigma : \mathfrak{a} \to \operatorname{End}_k(\mathfrak{r})$ on $\mathfrak{r}$ (by $\mathfrak{a}$) through derivations of $\mathfrak{r}$. This action of $\mathfrak{a}$ on $\mathfrak{r}$ extends naturally to an action $\sigma' : \mathfrak{a} \to \operatorname{End}_k(T(\mathfrak{r}))$ of $\mathfrak{a}$ on the tensor algebra $T(\mathfrak{r})$ of $\mathfrak{r}$, each $A \in \mathfrak{a}$ acting as a derivation of the associative algebra $T(\mathfrak{r})$. It is easily seen that $\sigma'(A)$ leaves invariant the left ideal $I_\mathfrak{r}$ generated by $\{X \otimes Y - Y \otimes X - [X, Y] | \ X, Y \in \mathfrak{r}\}$ in $U(\rho)$. Thus σ' defines a homomorphism of $\tilde{\sigma} : U(\mathfrak{a}) \to \operatorname{End}_k(U(\mathfrak{r}))$ such that $\tilde{\sigma}(U(\mathfrak{a}))(U_p(\mathfrak{r})) \subset U_p(\mathfrak{r})$ (and in particular $\tilde{\sigma}(U(\mathfrak{a}))(\mathfrak{r}) \subset \mathfrak{r}$). The left multiplication in $U(\mathfrak{r})$ defines a representation $L : U(\mathfrak{r}) \to \operatorname{End}_k(U(\mathfrak{r}))$: for $X \in \mathfrak{r}$ and $Z \in U(\mathfrak{r})$, $L(X)(Z) = X \cdot Z$. Combining the representations L of $\mathfrak{r}$ and $\tilde{\sigma}$ of $\mathfrak{a}$, we obtain a representation $\tau : \mathfrak{g} \to \operatorname{End}_k(U(\mathfrak{r}))$ as follows: for $(A, X) \in \mathfrak{a} \oplus \mathfrak{r} (\simeq \mathfrak{g}$ as a vector space) and $Z \in \mathfrak{g}$

$$\tau(A, X)(Z) = \sigma(A)(X \cdot Z).$$

It is an easy exercise to check that τ is indeed a representation of $\mathfrak{g}$. We will now prove the following theorem from which we will deduce Ado's theorem.

7.36. Theorem. *Assume that $\mathfrak{r}$ is solvable and let ρ be a finite dimensional representation of $\mathfrak{r}$ on a vector space V such that $\rho(\mathfrak{n})$ consists entirely of nilpotent matrices for a nilpotent ideal $\mathfrak{n}$ of $\mathfrak{g}$ contained in $\mathfrak{r}$ and containing $[\mathfrak{g}, \mathfrak{r}]$. Then there is a finite dimensional representation $\tilde{\rho}$ of $\mathfrak{g}$ such that ρ is a quotient of $\tilde{\rho}|_\mathfrak{r}$ and $\tilde{\rho}(\mathfrak{n})$ consists entirely of nilpotent endomorphisms of $V_{\tilde{\rho}}$.*

7.37. The proof of 7.35 will be achieved in a number of steps. For $X \in \mathfrak{r}$, let P_X denote the minimal polynomial of $\rho(X)_s$, the semi-simple part of $\rho(X)$. Let $\{X_i | \ 1 \le i \le r\}$ be a basis of $\mathfrak{r}$ such that $\{X_{p+j} | 1 \le j \le q\}$ where $p + q = r$ is a basis of $\mathfrak{n}$. Set $P_i = P_{X_i}$; then $P_i(T) = T$ for $i > p$. For an integer $l > 0$, let I_l denote the <u>left</u> ideal in $U(\mathfrak{r})$ generated by the set $\{P_{i_1}(X_{i_1}) \cdot P_{i_2}(X_{i_2}) \cdots P_{i_j}(X_{i_j}) \cdots P_{i_l}(X_{i_l}) | \ 1 \le i_j \le r\}$. The first step in the proof is the following.

7.38. Lemma. *For an integer $l \geq 1$, I_l is a 2-sided ideal in $U(\mathfrak{r})$, stable under $\tilde{\sigma}(\mathfrak{a})$.*

7.39. Proof. The lemma follows from the following.

Assertion: *Let $D : U(\mathfrak{r}) \to U(\mathfrak{r})$ be a derivation stabilizing $\mathfrak{r}$ and such that $D(\mathfrak{r}) \subset \mathfrak{n}$. Then I_l is stable under D.*

To see that the assertion implies the lemma, observe that for $X \in \mathfrak{r}$, the map $\tilde{ad}(X) : U(\mathfrak{r}) \to U(\mathfrak{r})$ defined by setting for $Z \in U(\mathfrak{r})$, $\tilde{ad}(X)(Z) = X \cdot Z - Z \cdot X$ is a derivation of $U(\mathfrak{r})$ stabilizing $\mathfrak{r}$ and inducing $ad_{\mathfrak{r}}(X)$ on it, so that $\tilde{ad}(X)(\mathfrak{r}) \subset \mathfrak{n}$. By the assertion, one has for $X \in \mathfrak{r}$ and $Z \in I_l$, $(X \cdot Z - Z \cdot X) \in I_l$. Since I_l is a left ideal, it follows that $Z \cdot X \in I_l$ for all $X \in \mathfrak{r}$. Since $\mathfrak{r}$ generates $U(\mathfrak{r})$ this means that I_l is a right ideal as well. For $A \in \mathfrak{a}$, $\tilde{\sigma}(A)$ is a derivation stabilizing $\mathfrak{r}$ and mapping $\mathfrak{r}$ into $\mathfrak{n}$. By the assertion I_l is stable under $\sigma(A)$ for every $A \in \mathfrak{a}$. Thus the lemma follows from the assertion.

7.40. Proof of the Assertion. We argue by induction on l. When $l = 1$, we need to show that if D is a derivation as in the assertion then, $D(I_1) \subset I_1$. Now an element of I_1 is of the form $\sum_{j=0}^{r} Z_j \cdot P_j(X_j)$ with $Z_j \in U(\mathfrak{r})$ and hence it suffices to show that

$$D(Z_j \cdot P_j(X_j)) = D(Z_j) \cdot P_j(X_j) + Z_j \cdot D(P_j(X_j)) \qquad (*)$$

is in I_1. As the first term on the right hand side of $(*)$ is evidently in I_1, we need to only show that for any derivation D with $D(\mathfrak{r}) \subset \mathfrak{n}$, $D(P_j(X_j)) \in I_1$. Now $P_j(X_j) = \sum_{i=0}^{n_j} a_{j,i} \cdot X_j^i$; so we need only show that $D(X_j^t) \in I_1$ for all integers $t \geq 1$ (for a derivation D with $D(\mathfrak{r}) \subset \mathfrak{n}$). This is clear when $t = 1$. Assume that $D(X_j^i) \in I_1$ for $i \leq t$ (for all derivations D of $\mathfrak{r}$ taking $\mathfrak{r}$ into $\mathfrak{n}$). Then one has

$$D(X_j^{t+1}) = X_j \cdot D(X_j^t) + D(X_j) \cdot X_j^t$$

$$= X_j \cdot D(X_j^t) + X_j^t \cdot D(X_j) + \tilde{ad}(D(X_j))(X_j^t).$$

By the induction hypothesis the first two terms on the right hand side of the above equation are in I_1. On the other hand since $ad(D(X_j))$ is a derivation taking $\mathfrak{r}$ into $\mathfrak{n}$, again by the induction hypothesis, the third term is in I_1 as well, proving the assertion when $l = 1$. In particular this proves that I_1 is a 2-sided ideal.

7.41. Suppose then that we have proved that the assertion holds for all integers $l < s$. For an ordered s-tuple of positive integers $J = \{J(t) \in [1, r]| \; 1 \leq t \leq s\}$ set $Z_J = \prod_{1 \neq t \leq s} P_{J(t)}(X_{J(t)})$ (the product taken in the increasing order of t). Then $\{Z_J| \; J : [1, s] \to [1, r]\}$ generate the left ideal I_s. Set $Z'_J = \prod_{1 \neq t < s} P_{J(t)}(X_{J(t)})$, so that $Z_J = Z'_J \cdot P_{J(s)}(X_{J(s)})$. One then has $D(Z_J) = D(Z'_J) \cdot P_{J(s)} + Z'_J \cdot D((P_{J(s)})(X_{J(s)}))$. As $Z'_J \in I_{r-1}$, by the induction hypothesis, the $D(Z'_J) \in I_{s-1}$, and $D((P_{J(s)})(X_{J(s)})) \in I_1$. I_{s-1} and I_1 are moreover (by induction hypothesis) 2-sided ideals. It follows that

$D(Z_J) \in I_{s-1} \cdot I_1 = I_s$ proving that the Assertion holds for s. This completes the proof of the assertion and hence the lemma.

7.42. Corollary to Assertion. *For $A \in \mathfrak{a}$, $\sigma(\tilde{A})(I_n) \subset I_n$.*

7.43. Lemma. *I_l has finite co-dimension in $U(\mathfrak{r})$ for all $l > 0$.*

7.44. Proof. We will show that the $\{X^\alpha | \ |\alpha| \leq l \cdot r \cdot a\}$ where $a = \max\{degree$ $(P_i)| \ 1 \leq i \leq r\}$ span $U(\mathfrak{g})$ modulo I_l. Let E be the span of I_l and $\{X^\alpha | \ |\alpha| \leq l \cdot r \cdot a\}$. It is clear that $U_{l \cdot r \cdot a}(\mathfrak{r}) \subset E$. Suppose that we have proved that for some $N \geq l \cdot r \cdot a$, $U_N(\mathfrak{r}) \subset E$. Let $\beta \in \mathbb{N}^r$ be such that $|\beta| = N + 1$, i.e., $\sum_{i=1}^r \beta_i = N + 1$. Then there is an i such that $\beta_i > l \cdot a$. Now $X^\beta - (X_i^{\beta_i} \cdot X^{\beta - \beta_i \cdot <i>}) \in U_N(\mathfrak{r}) \subset E$. Hence it follows that we need only show that $X_i^{\beta_i} \cdot X^{\beta - \beta_i \cdot <i>} \in E$. Set $\beta' = \beta - \beta_i \cdot <i>$. Now $X_i^{\beta_i} = Q_i(X_i) \cdot P_i(X_i)^l + R_i(X_i)$ with R_i a polynomial of degree $l \cdot e_i < l \cdot a$ so that $R_i(X_i) \cdot X^{\beta'} \in U_N(\mathfrak{g}) \subset E$. Thus we need to only show that $Q_i(X_i) \cdot P_i(X_i)^l \cdot X^\beta \in E$ and this is obvious since $P_i(X_i)^l \in I_l$ and I_l is a 2-sided ideal.

7.45. Proof of Theorem 7.35. We saw in 7.34 that there is a natural representation τ of $\mathfrak{g}$ on $U(\mathfrak{r})$ with $\tau(X)$ for $X \in \mathfrak{r}$ being the left multiplication by X. From 7.37, we see that I_l is a $\mathfrak{g}$-submodule of $U(\mathfrak{r})$. Let $\tilde{\rho}$ be the finite dimensional representation of $\mathfrak{g}$ (as well as $U(\mathfrak{g})$) on $U(\mathfrak{r})/I_l$. Now the kernel of $\rho = \tilde{\rho}|_{U(\mathfrak{r})} : U(\mathfrak{r}) \to \mathrm{End}_k(V_\rho)$ contains I_l and thus $\rho(U(\mathfrak{r}))$ is a $\mathfrak{r}$-module (denoted ρ^*) which is a quotient of $\tilde{\rho}|_{U(\mathfrak{r})}$. If $\{v_i| \ 1 \leq j \leq d\}$ is a basis of V_ρ, V_ρ is the sum of the $\mathfrak{r}$-submodules $\{U(\mathfrak{r}) \cdot v_i| \ 1 \leq i \leq d\}$. It follows that V_ρ is the quotient of the direct sum of d copies of ρ^*; and ρ^* is a quotient of $\tilde{\rho}|_{\mathfrak{r}}$. Thus ρ is a quotient of a direct sum of d copies of $\tilde{\rho}|_{U(\mathfrak{r})}$. Finally since $X^l \in I_l$ for every $X \in \mathfrak{n}$, we see that $\tilde{\rho}(X)^l = 0$. Thus $X \in \mathfrak{n}$ maps to a nilpotent element in the direct sum of d copies of $\tilde{\rho}$. This completes the proof of the theorem.

7.46. Corollary. *The notation is as in Theorem 7.35. Then given a finite dimensional representation ρ of $\mathfrak{r}$ with $\rho(X)$ nilpotent for every $X \in \mathfrak{n}$, there is a finite dimensional representation $\bar{\rho}$ of $\mathfrak{g}$ such that $\bar{\rho}(X)$ is nilpotent for every $X \in \mathfrak{n}$ and ρ is isomorphic to a $\mathfrak{r}$-sub-representation of $\bar{\rho}$.*

This is immediate from applying Theorem 7.35 to the dual of ρ.

7.47. Proof of Ado's Theorem: The Case of Nilpotent $\mathfrak{g}$. In this case we will prove a stronger statement viz., that $\mathfrak{g}$ admits a faithful representation ρ such that $\rho(X)$ is nilpotent for every $X \in \mathfrak{g}$. We will prove the theorem by induction on $\dim_k \mathfrak{g}$. Now $\mathfrak{g}$ admits a nilpotent ideal $\mathfrak{n}$ of co-dimension 1. By the induction hypothesis, $\mathfrak{n}$ admits a faithful finite dimensional representation ρ' such that $\rho'(\mathfrak{n})$ consists entirely of nilpotent matrices. Evidently, $\mathfrak{g}$ is the semi-direct product of $k \cdot X$ and $\mathfrak{n}$ where $X \notin \mathfrak{n}$. By 7.45, there is a finite dimensional representation ρ of $\mathfrak{g}$ such that $\rho(\mathfrak{n})$ consists entirely of nilpotent matrices and contains ρ' as a sub-representation. Define the representation $\sigma : \mathfrak{g} \to V_\rho$ by setting $\sigma(Z) = \rho(Z)$ for $z \in \mathfrak{n}$ and $\sigma(X) = \rho(X)_n$, the nilpotent part of $\rho(X)$. It is easily checked that σ is indeed a representation of $\mathfrak{g}$ and that $\sigma(\mathfrak{g})$ consists entirely of nilpotent endomorphisms of $V_\sigma (= V_\rho)$. The quotient $\mathfrak{h} = \mathfrak{g}/\mathfrak{n} (\simeq k \cdot X)$ is of dimension 1 and hence admits a faithful representation ν on k^2 with $\nu(X)$ being a non-zero upper triangular nilpotent matrix for a $0 \neq X \in \mathfrak{h}$. Treating ν as a representation of $\mathfrak{g}$ we see that $\mu = \nu \oplus \sigma$ is a faithful representation of $\mathfrak{g}$. It is clear that $\mu(\mathfrak{g})$ consists entirely of nilpotent matrices.

7.48. Proof of Ado's Theorem: The Case of Solvable $\mathfrak{g}$. We argue by induction on $\nu(\mathfrak{g}) = \dim_k(\mathfrak{g}/\mathfrak{n})$, where $\mathfrak{n}$ is the maximal nilpotent ideal in $\mathfrak{g}$. When $\nu(\mathfrak{g}) = 0$, $\mathfrak{g}$ is nilpotent and the result is already established. Now $\mathfrak{g}$ is the semi-direct product $k \cdot X$ and a co-dimension 1 ideal $\mathfrak{g}'$ containing $\mathfrak{n}$. Suppose that ρ is a faithful representation of $\mathfrak{g}'$ such that $\rho(\mathfrak{n})$ consists entirely of nilpotent matrices – such a ρ exists by induction hypothesis. By 7.45, there is a representation $\tilde{\rho}$ of $\mathfrak{g}$ such that ρ is a sub-representation of $\tilde{\rho}|_{\mathfrak{g}'}$ and $\tilde{\rho}(\mathfrak{n})$ consist entirely of nilpotent matrices. Let ν be a faithful finite dimensional representation of the 1-dimensional Lie algebra $\mathfrak{g}/\mathfrak{g}'$; then $\mu = \nu \oplus \tilde{\rho}$ is a faithful finite dimensional representation of $\mathfrak{g}$ with $\mu(\mathfrak{n})$ consisting of nilpotent endomorphisms.

7.49. Proof of Ado's Theorem: The General Case. We know that any Lie algebra $\mathfrak{g}$ is the semi-direct product of a semi-simple subalgebra $\mathfrak{a}$ and the (solvable) radical $\mathfrak{r}$. Also if $\mathfrak{n}(\mathfrak{g})$ is the maximum normal nilpotent ideal of $\mathfrak{g}$, $[\mathfrak{g}, \mathfrak{r}] \subset \mathfrak{n}(\mathfrak{g})$. Let ρ be a faithful finite dimensional representation of $\mathfrak{r}$ such that $\rho(\mathfrak{n}(\mathfrak{g}))$ consists of nilpotent endomorphisms. By 7.45, there is a faithful finite dimensional representation $\tilde{\rho}$ of $\mathfrak{g}$ such that ρ is a sub-representation of $\tilde{\rho}|_{\mathfrak{r}}$ and $\tilde{\rho}(\mathfrak{n}(\mathfrak{g}))$ is contained in the set of nilpotent endomorphisms of $V_{\tilde{\rho}}$. Let ν denote the adjoint representation of $\mathfrak{g}/\mathfrak{r} (\simeq \mathfrak{a})$ considered as a representation of $\mathfrak{g}$. Since $ad_{\mathfrak{a}}$ is a faithful representation, $\mu = \nu \oplus \tilde{\rho}$ is a faithful representation of $\mathfrak{g}$. This completes the proof of Ado's theorem.

Applying Lie's theorem on the correspondence between Lie subalgebras of the Lie algebra $\mathfrak{g}$ of the Lie group G over a local field k and Lie subgroups of G, we conclude that we have the following.

7.50. Theorem. *Let k be a local field and $\mathfrak{g}$ a Lie algebra over k. Then there is a Lie group G over k such that the Lie algebra of G is isomorphic to $\mathfrak{g}$.*

7.51. Corollary. *Let k be an archimedean local field. The assignment to each simply connected Lie group G over k, its Lie algebra $\mathfrak{g}$ is an equivalence of the category of simply connected Lie groups over k and the category of Lie algebras over k.*

7.52. A more comprehensive structure theory would include a classification of a semi-simple Lie algebras over k as well as their representation theory. This would involve a classification (and representation theory) over an algebraically closed field first, and then obtain results over arbitrary fields using Galois Cohomology. An excellent account of the classification and representation of Lie algebras over an algebraically closed field is to be found in Serre [13]. The classification over $\mathbb{R}$ is dealt with in Helgason [4]. A quick introduction to (non-abelian) Galois Cohomology is to be found in Springer [16]. The classification over p-adic fields is described in Tits [17].

References

(1) M Artin, Algebra, Prentice Hall, 1991.

(2) N Bourbaki, Groupes et algèbres de Lie Chapitres 1, 2 et 3, Paris, Hermann, 1971, 1973, 2007.

(3) Allen Hatcher, Algebraic Topology, https://pi.math.cornell.edu >~ hatcher PDF

(4) Sigurdur Helgason, Differential Geometry, Lie Groups, and Symmetric Spaces, Academic Press, 1978.

(5) N Jacobson, Basic Algebra (volumes I and II), Hindustan Publishing Corporation, 1983.

(6) John L Kelley, General Topology: 27 (Graduate Texts in Mathematics), Springer Verlag, 1975.

(7) James Munkres, Topology, a First Course, Prentice Hall, 1974.

(8) R Narasimhan, Topics in Analysis, TIFR Lecture Notes (Free download available in TIFR web-site).

(9) L Pontrjagin, Topological Groups (translated from Russian by Emma Lehmer), Princeton Mathematical Series (avialable for free download at https://archive.org/details/in.ernet.dli.2015. 89986/page/n5/mode/2up).

(10) Alain M Robert, A Course in p-adic Analysis: 198 (Graduate Texts in Mathematics), Springer Verlag, 2000.

(11) W Rudin, Principles of Mathematical Analysis (Indian Edition), 3rd ed., McGraw-Hill, 1976.

(12) Jean-Pierre Serre, Local Fields. Graduate Texts in Mathematics. Vol. 67, Springer Verlag, New York Berlin Heidelberg, 1979

(13) Jean-Pierre Serre, Complex Semisimple Lie Algebras, Springer Monographs in Mathematics, 2001.

(14) Jean-Pierre Serre, Lie Algebras and Lie Groups, Lecture Notes in Mathematics, Vol 1500, Springer-Verlag, Berlin Heidelberg, 1992.

(15) Michael Spivak, Calculus on Manifolds, A Modern Approach to Classical Theorems of Advanced Calculus, Westview Press, 1965.

(16) T A Springer, Galois Cohomology of Linear Algebraic Groups, 'Algebraic Groups and Discontinuous Groups', Proc. Symp., Pure Math 33, Vol 9, 149–158 American Mathematical Society, 1966.

(17) J Tits, Reductive groups over local fields, Automorphic forms, representations and L-functions, Proc. Symp., Pure Math 33, Part 1, Corvallis, 1979.

(18) A Weil, Basic Number Theory, Vol 144, Grundlehren der mathematischen Wissenschaften Springer Verlag, 1973.

Index

In this index, only the first occurrence of the word is indicated except when a second occurrence has a different context.

The Alphabet in Roman and Gothic Scripts

A 𝔄 a 𝔞   B 𝔅 b 𝔟   C ℭ c 𝔠

D 𝔇 d 𝔡   E 𝔈 e 𝔢   F 𝔉 f 𝔣

G 𝔊 g 𝔤   H ℌ h 𝔥   I ℑ i i

J 𝔍 j j   K 𝔎 k 𝔨   L 𝔏 l l

M 𝔐 m 𝔪   N 𝔑 n 𝔫   O 𝔒 o o

P 𝔓 p p   Q 𝔔 q q   R ℜ r 𝔯

S 𝔖 s 𝔰   T 𝔗 t t   U 𝔘 u u

V 𝔙 v 𝔳   W 𝔚 w 𝔴   X 𝔛 x 𝔵

Y 𝔜 y 𝔶   Z 𝔷 z 𝔷

Texts and Readings in Mathematics